Digital Proxemics

Steve Jones
General Editor

Vol. 110

The Digital Formations series is part of the Peter Lang Media and Communication list.
Every volume is peer reviewed and meets
the highest quality standards for content and production.

PETER LANG
New York • Bern • Frankfurt • Berlin
Brussels • Vienna • Oxford • Warsaw

John A. McArthur

Digital Proxemics

How Technology Shapes
the Ways We Move

PETER LANG
New York • Bern • Frankfurt • Berlin
Brussels • Vienna • Oxford • Warsaw

Library of Congress Cataloging-in-Publication Data

Names: McArthur, John A., author.
Title: Digital proxemics: how technology shapes the ways we move / John A. McArthur.
Description: New York: Peter Lang, 2016.
Series: Digital formations; vol. 110 | ISSN 1526-3169
Includes bibliographical references and index.
Identifiers: LCCN 2015036105 | ISBN 978-1-4331-3187-5 (hardcover: alk. paper)
ISBN 978-1-4331-3186-8 (paperback: alk. paper) | ISBN 978-1-4539-1724-4 (e-book)
Subjects: LCSH: Mobile computing. | Mobile computing—Social aspects.
Location-based services. | Near-field communication.
Classification: LCC QA76.59.M43 2016 | DDC 004—dc23
LC record available at http://lccn.loc.gov/2015036105

Bibliographic information published by **Die Deutsche Nationalbibliothek**.
Die Deutsche Nationalbibliothek lists this publication in the "Deutsche
Nationalbibliografie"; detailed bibliographic data are available
on the Internet at http://dnb.d-nb.de/.

© 2016 Peter Lang Publishing, Inc., New York
29 Broadway, 18th floor, New York, NY 10006
www.peterlang.com

TABLE OF CONTENTS

PREFACE

When people ask me what I study, I often reply, "Proxemics. The ways we use space." What fascinates me is that every single time I offer this response, people further that idea based on their own unique experiences of space. They tell me stories about the importance of space in their lives. Some reference interpersonal distance, geography, or gesture. Some describe physical environments or places of particular significance to them. Others ask if I mean interior design, or architecture, or public performance. And still others, the science fiction buffs, immediately inquire about my political leanings: *Star Trek* or *Star Wars*? For the record, it's *Star Wars*. The study of proxemics can investigate all of these things, but it is fundamentally focused on the behaviors of humans in relation to one another as they use and experience space. Each person has his own view of the ways that space is used in his life, and his own lens for understanding why space matters. The point is that all people feel a connection to space and the ways they use space, even if that connection exists under the surface of cognition, behind the walls of cultural understanding. Proxemics seems obvious when it is described to us and yet the ways we use proxemics every day exists beyond our awareness.

In 1966, Edward Hall coined the term 'proxemics' (pronounced präk-SEE-miks) in his book *The Hidden Dimension*[1] to name an as-yet-undocumented aspect

VIII   DIGITAL PROXEMICS: HOW TECHNOLOGY SHAPES THE WAYS WE MOVE

of our lived experience. His studies of human interactions across cultures led him to describe proxemics as the study of human use of space in personal, social, and public contexts. Now, leading up to the 50th anniversary of the publication of that book, proxemics has been applied widely in business, cultural, social, and educational settings as an important component of nonverbal communication, sociology, psychology, and cultural critique. However, over that half century, the concept of proxemics has remained static while the cultural landscape has shifted, due in large part to the rise of digital communication technologies. This book seeks to examine and reflect upon the role of digital communication in proxemics, offering explorations of the ways digital technology shapes our personal bodily movement, our interpersonal negotiation of social space, and our navigation of public spaces and places. The result is both a spatial turn in the study of digital technology and a digital turn in the study of proxemics.

Digital Proxemics incorporates three themes to explore the human-technology interaction in proxemics: (1) the mutual impacts of us on space and space on us; (2) the lens of experience design theory as a framework for analysis; and (3) the interaction between the digital and the physical in the spaces we inhabit. These three ideas create a foundation for a conversation about how technologies, spaces, and people mutually act upon each other as we experience environments. Whereas much writing on the subject of digital technology and spaces focuses on what digital technology might do *for us*, this book explores what the same technology might do *to us*. In addition, it explores the ways we might respond to this intersection through reflection and digital literacies.

Rather than writing in the form of a traditional academic research monograph, I use narratives and stories throughout the book to illustrate the concepts as a reminder that studies of proxemics are studies of self. This focus on proxemics as a window to self and individual culture echoes the tone of Edward Hall's work on proxemics in the 1960s. In the popular conversation surrounding proxemics, Hall's text remains the most cited, most referenced work on proxemics. *Digital Proxemics* is the first text that attempts to refocus proxemics in a digital world, although this type of work is happening among other facets of verbal and nonverbal communication. As such, this concept of digital proxemics will become an enhanced foundation for studying the interaction between human communication and technology surrounding the concept of proxemics. In addition, the study of digital proxemics attempts to reintroduce proxemics for a new generation of scholars, researchers, and citizens as a marker of our lived experience.

In popular circles, the widespread popularity of the 99% *invisible* podcast, Eli Pariser's *The Filter Bubble*,[2] Sherry Turkle's *Alone Together*,[3] and many other works of cultural reflection speak to the excitement surrounding this kind of text. Howard Rheingold's *Smart Mobs*[4] discusses the coming connection between mobile technology and users, but veers away from proxemics as a site for cultural shift. In academic circles, research monographs focus on the relationship between mobile technology and public spaces (e.g., de Sousa e Silva & Frith's *Mobile Interfaces in Public Spaces*[5] and Jason Farman's *Mobile Interface Theory*[6]), yet none discuss the impacts of technology on our understanding of proxemics. This book seeks to enter the conversation at the intersection of the authors mentioned above and others, by providing relatable evidence of cultural shift framed in the context of physical and digital spaces.

Readers of this book might approach it from several vantage points. Students of communication, sociology, anthropology, and related fields should find that this text furthers contemporary thinking about proxemics as we have known it. For these readers, I hope the book functions as a vehicle for digital communication to enter the conversation as well as a resource for connecting with interdisciplinary research and thinking across a wide variety of technologically savvy ideas. Conversely, developers of mobile computing, ubicomp, and digital application technologies should find a call to consider not only what amazing things technology can do in spaces, but also how these technologies might privilege or diminish particular human actions in those spaces. For these designers, my hope is that the book causes us to evaluate technological progress not only by what we create but also how our creations inspire citizenship, community, and conversation. Researchers and thinkers in human communication and technology interaction should find that this text bridges a set of distinct fields that should (and sometimes do) learn from each other. I hope these researchers find proxemics to be a useful construct for integrating the user, ethical quandary, and agency into our ongoing discussions surrounding technology and the human experience. And many readers will evaluate this book as citizens of a rapidly changing and increasingly digitized culture. I hope that all of us will benefit from making visible our uses of space and that we can spend time pondering the importance of today's choices for the future of digital technologies in the spaces we inhabit.

The chapters in this book examine the ways that our uses of physical and digital spaces and our uses of technology are converging. Chapter one explores our daily interactions with spaces and technologies as actors in our environments, calling on us to reflect on the roles we allow digital technologies to play in our use of space. The chapter makes the case that we need moments of pause that invite us all to learn about the ways our culture—and our use of space as a marker of culture—is changing in response to digital technologies.

Through an exploration of Edward Hall's and others' works on proxemics, chapter two describes major components that ground the study of proxemics, including human distances, territories, and lived experiences in spaces. Then, the chapter explores the cultural shifts that necessitate a furthering of the concepts of proxemics for a digital society. The result is a theory of digital proxemics in which technology allows spaces and users through four actions, giving each the ability to capture, inform, alter, and control the other.

Chapter three applies the concepts of digital proxemics to the flagship notion in proxemics: the interpersonal distancing of the body in relation to others. In this conversation, the four zones of interpersonal distance are envisioned in the context of digital gathering. In addition, this chapter discusses the sociospatial functions of digital technology as both a sociofugal and a sociopetal force.

The first in a series of chapters that contextualizes digital proxemics in specific settings, chapter four focuses on the human body and the ways digital technologies change the physical motion of the body. Digital technologies not only capture, digitize, and alter bodily motion, but they also can function as extensions and extenders of the human body. Opportunities for body-technology connections are discussed, including postural variation, motion capture devices, rotoscoping, prosthetics, and bionics.

Beginning with architect Jan Gehl's assertion that cities should be designed for people, chapter five examines the impacts of digital proxemics in communities, cities, and geographic locations. The chapter also introduces a new concept of technologies as facilitators of proxemofugal or proxemopetal experiences. By exploring the mapping and re-mapping of environments, sentient objects, and augmented reality, chapter five reflects on our agency—that is our ability to make our own choices—in the navigation of environments, which remains a fundamental component of our connection to places.

Chapter six introduces the concept of the archive as a component of space and place. Significant places for people are characterized by their connections to historical and/or emotional constructs surrounding a location. Spaces locate us in a variety of ways, and this chapter posits that people, digital technologies, and environments all create distinct archives of place that document our movements. Then, the chapter reflects on the relationship among archives, privacy, and surveillance.

Digital proxemics are contextualized in chapter seven through spaces once forbidden to visitors but now accessible via digital technology, as well as spaces created totally in virtual environments. Physical spaces addressed include spaces too small or large to be seen with the human eye, spaces too remote to physically encounter, spaces too dangerous to visit, and private events shared publicly. This chapter also addresses virtual environments as fabricated spaces and places developed using 3D immersive display (virtual reality) devices.

Chapter eight addresses the role of digital literacies in helping us understand and contribute to our lived experience with and through digital technology. Building on common approaches to digital and media literacy, this chapter seeks to complicate our understanding of digital proxemics by positioning users as actors in the spaces we visit and inhabit. Actors, unlike users, possess the ability to be decisive in ways that rewrite and reinscribe proxemic experiences.

Chapter nine investigates the research of proxemics, starting with Edward Hall's assertions about the study of proxemics as a study of self. First, it compiles and describes the constructs of proxemics: movement of the body, features and territories, interpersonal distance, sociospatial design, and cultural constructs of space. Then, the chapter identifies and explores potential research methodologies used in a variety of the articles and studies presented in the text, including observation, experimental designs, applied research big data, and criticism.

For readers unfamiliar with information design and user-experience design, the appendix introduces experience design as a lens for the study of space, calling on readers to recognize the designed elements of the spaces they visit and inhabit. A synopsis of the field of information design culminates in the shift in thinking from designing information to designing experiences. This shift (reflected in the work of Saul Carliner, Nathan Shedroff, Donald Norman, B. J. Fogg, Maya Lin, and Patrick Jordan, among others) is a result of a focus on the user rather than a focus on content.

Overall, this book is intended to be an exploration of the intersection of physical space and digital technologies. We find ourselves standing at that intersection determining which ways we will move, and the decisions we make today will have repercussions for our future. As I hope you'll agree, the decisions about which ways we move remain ours to make, for now. But, as our culture changes, our ability to make choices about how to move will also be called into question, as will our expectations for what roles technology will play in our lives. As we navigate this intersection together, I hope *Digital Proxemics: How Technology Shapes the Ways We Move* will be at once a valuable lens through which we can view our shifting culture, a cautionary tale through which we might envision problematic outcomes, and an optimistic projection of possibility for the future of human communication and technology interaction.

John A. McArthur
October 2015

ACKNOWLEDGMENTS

When I started this book, I was new to the book publishing process. I'm still new, but I am grateful to have had much support along the way. Dr. Andy Billings, ever the friend, mentor, and academic standard-bearer, guided me toward multiple paths to pursue. I continue to be grateful that our lives intertwined at Clemson. Vanessa Domine, Mohammed el-Nawawy, and Nancy Clare Morgan clarified early questions about the publishing process for me. Mary Savigar helped me navigate the intersection of my interests with the needs of the academic community. Sophie Appel designed the striking cover art and Bernie Shade led me through the design phases. Digital Formations editor Steve Jones provided counsel and feedback.

Several colleagues and friends in my personal learning network volunteered to read chapters of this book as I wrote them. They have offered me advice, suggested new lines of thought, and helped me clarify ideas and approaches to the information presented. Thank you to Dr. Zachary White, Professor Ashley Fitch Blair, and Dr. Corinne Weisgerber. Other colleagues remain anonymous reviewers of papers I've submitted to a variety of journals and conferences on the topics herein. Still others have responded to the blog posts, tweets, social media updates, and chats that have served as drafts of these ideas. They have all improved the book in countless ways and helped seek out new facts, fresh ideas, and innovative concepts. The errors that persist in the book are all mine.

I am thankful for the graduate program in the James L. Knight School of Communication at Queens University of Charlotte and for my colleagues who have worked for its success: Kim, Zachary, Leanne, Alexis, Daina, and Van. I continue to be inspired by their creativity. The innovative thinking behind the program allowed me to investigate ideas that honed my thinking about space, place, and digital technology. The partnerships I've had with graduate students as co-researchers and sounding boards has enhanced my research and my thinking, and challenged me to think in new ways about new spaces. In particular, I was motivated by the intellectual curiosity and research interests of my graduate students Kristen Bostedo-Conway, Valerie Johnson Graham, Ashleigh Farley White, Jennifer Hull, Tamara Blue, Genevieve Bland, Morgan Lloyd, June Furr and all the students who participated in my graduate seminar in Space, Technology, and Information Design. Their excellent investigations into space and technology have colored and shaped my thinking and pointed me in new directions.

Last, and most important, I am grateful for Erin. She gave me the space and time I needed to think, write, and edit. In addition to charting my progress on our wall, she made this book a family goal by inviting Eleanor to write books of her own, caring for our newborn daughter Violet, and teaching us all how to celebrate even the smallest amounts of success along the way. I know it was not easy on her, but her generosity, support, and encouragement made all the difference.

· 1 ·

MOVING

In December 1988, my parents purchased the penultimate present for my brothers and I—the Nintendo Entertainment System (NES). I remember the excitement of the morning. We discovered NES sitting atop the television, or more specifically atop the 4-foot-wide wooden cube that housed the television and occupied one corner of our den. The oak grain of the television juxtaposed with the sleek black and gray of the NES console foreshadowed the rapidly changing future of home entertainment. This obvious nod to the future was all but lost on three boys.

The now iconic NES game cartridges had futuristic, golden connectors protruding from the bottom that merged the game to the console. When these connectors slid into the console, any dust that had accumulated on them disrupted the images on the screen and the play of the game. The fix for this bug was to remove the cartridge and blow mightily on the connectors in hopes of removing the dust. Even today, when I see people my age blowing dust out of their iPhones, I can only assume that this motion is a holdover of Nintendo's influence on my generation.

With the release of NES, Nintendo attempted to reinvigorate the gaming market following the video game crash of 1983.[1] The crash, resulting from market saturation of a variety of consoles and poorly designed games, left an opening for a single system to set industry standards. NES was released in America through limited testing in 1985 and then mass marketed in 1986 with widespread appeal. With a whopping 8-bit resolution, the graphics of NES outpaced those of Atari's *Frogger* and other replications of popular arcade games. Because of its role in the resurgence of video gaming, NES was named the best video game system of all time in 2012 by IGN Entertainment.[2]

By 1988, my brothers and I were just slightly behind the gaming curve, but that was undoubtedly the year to get in the game. In the category of technology's impact on movement, 1988 was the system's banner year. That year, NES was sold with the "Power Set," which included the console, *Super Mario Bros.* with two controllers, *Duck Hunt* with the NES Zapper, and *World Class Track Meet* with the PowerPad.

Super Mario Bros. popularized what we now might consider traditional video gaming. Mario and Luigi, avatars on-screen, responded to player instructions via buttons on a controller. At the time, my brothers and I didn't know that the bright red buttons on the controllers would prepare our thumbs for the onslaught of texting that would follow decades later. My parents and other adults warned us against injuries to our digits, but those thumbs experienced growth in strength and dexterity. In the process, the Super Mario Brothers taught the McArthur brothers about the domains of video games.[3] While our hands operated the controllers manually, the space of play was fully on-screen. Mario and Luigi were engaged in a mushroom and turtle-filled adventure that occurred inside our television. That spatial dynamic was turned upside down with PowerPad.

I was never much of a track and field athlete—until PowerPad. The plastic floor mat of PowerPad was embedded with 12 pressure sensors connected by a cord to the NES console. The square design of the mat bore 12 "buttons" with 6 blue buttons arranged on the left and 6 red buttons arranged on the right. Two players could compete with each other, in turn, by standing on the two sides of the mat. The space of play became our den, and the avatar on screen became secondary.

In a traditional track and field contest, the long jump is a test of an athlete's ability to run and leap into the air, landing in a pit of sand as far as possible from the starting line. However, in *World Class Track Meet*, the long jump was a test of precision. Savvy children (and I give full credit for this discovery in our house to my younger brothers) quickly learned that PowerPad only registered movement that occurred on its sensors. When I watched my brother David compete in the long jump, his actions went something like this: (1) Run on tiptoes as fast as possible. (2) When the avatar approaches the line, leap to the side off the PowerPad. (3) Fidget while trying to wait patiently. (4) Firmly touch right big toe to PowerPad sensor to land. The third component was the trickiest. Touch too soon and David would be out-jumped by the next player. Touch too late and the avatar slid through a cloud of dust as his face planted into the sand. The precision was in the timing of moving back into the space.

In 1989, I referred to the actions described above as "cheating." Today, I fully recognize that NES and PowerPad reinscribed the movement of the long jump.

Delighted at the PowerPad's ability to help three boys work up a sweat in the winter, my parents quickly invested in *Super Team Games*. *Super Team Games* included the crab walk, log jump, and tug-of-war. We quickly learned the best ways to perform these movements on the PowerPad for maximum points, despite the fact that these motions, like that of the long jump, seldom resembled the games played on screen. PowerPad did something that regular games had not—it got us moving.

Looking back, PowerPad was the first motion-capture device I remember using. Later generations might look at Nintendo's Wii Fit, Dance Dance Revolution, or other exergaming[4] systems in the same light. In my own reflections on the impact of digital tools on physical spaces, the PowerPad and its long jump will always serve as a benchmark for the interaction of technology and user.

I don't want to give the wrong impression that this is a book about gaming. It's not. In video games, the user manipulates a physical device in conjunction with sensorial feedback (visual and auditory) on screen. Players' movements are captured and encoded to control the movements of an avatar in digital space. Video games give us a valuable window into the intersections of user, technology, and space. But rather than the digital spaces of video games, this book is focused on the physical spaces of the world.

This work presupposes that digital technology could impact our movements in the physical world in dynamic ways, some akin to ways we move avatars in video games, and some quite different. The single example above, PowerPad's long jump, offers a simple illustration of one way that digital technology rewrites movement, while controller manipulation, exercise as performance, and the act of blowing air on a game cartridge also serve as inscriptions of motion in space. This book seeks to investigate various ways we move in spaces and to reflect upon the multiple roles of digital technologies in our use of space.

Digital technology shapes the ways we move.

Actors in Environments

Movement has many facets. We move our bodies. When I accept a new job, my family moves. On a trip, we move in, around, and through cities. I can be moved by a book or my relationship with a loved one. We are bodies in motion, moving through spaces, being moved by experiences. Each form of movement highlights at least one interaction we have with space. Our bodies take up space. We move our bodies in and through spaces as we exist, explore, and experience them. Spaces surround us. We visit, remember, and reminisce about places. Spaces are at once locational, experiential, visceral, and emotional. Our interactions with space are also interactions with self.

When Edward Hall coined the term *proxemics* in *The Hidden Dimension* (1966), he used it to mean "social and personal space and man's perception of it."[5] The role of space in our everyday lives is part of our culture. Our learned behaviors, he argues, are as foundational to our use of space as they are to other forms of verbal and nonverbal communication. In Hall's terms, the ways that we move our bodies in space and negotiate interpersonal distances with others are markers of culture. His contemporary, Robert Sommer, took these claims one step further by researching human use of built spaces. After multiple studies, Sommer's conclusion was that people are hindered by our "fatalistic acceptance" of the spaces we inhabit.[6] When considering both of these researchers together, we might conclude that interactions between humans and interactions of humans with and in spaces are products of culture and lived experience, influenced in compelling ways by the design of the spaces we inhabit.

In a graduate seminar last year, I made the bold claim that all experiences of space are designed for us. One student in the seminar rolled her eyes and, face aghast, said, "I don't believe that." As any good professor would do, I turned her sentiment into a full-length class discussion. After I inquired about her statement, she relayed a story of a hike she took with others through an undeveloped wood to view a beautiful waterfall. In her mind, the forest was as natural a setting as she'd ever encountered. Surely, she noted, that experience was not designed. We, I and the other members of the class, listened as she described the undisturbed beauty of the waterfall, the trees overhanging the serene river at the bottom of the falls, and the rocky hike she traversed to glimpse such natural wonders. Like her, I too revel in the joy and beauty of natural settings and quickly voiced my appreciation for mountain vistas. Then, I apologized to her in advance of my next statements.

Your experience was surely designed for you, I said. I offered her a list of questions. Did you follow a trail to get there? Did the trail lead directly to the most picturesque view of the falls? Did signage direct you to the trailhead or through the hike? Were the wood and waterfall located in a national, state, or local park? Or was the land conserved by a named or anonymous benefactor? Was the water flowing from or into a reservoir? Was the site clean and free of debris? Were toilets, postcards, or donation boxes present in the vicinity? Affirmative answers to any of these questions would point to an experience that had been designed. Even the act of selecting natural locations to conserve and preserve among a series of developed locations is an intentional—and I would argue important and noble—act of design. That discussion may have changed the way she experienced space. Over the next weeks, she looked everywhere for (and regularly discovered) indicators that her experiences may, in fact, be designed.

Without critical reflection, we would never realize that our experiences of spaces are designed for us. Like my student, I have lost the ability to simply visit a space. As soon as I walk into an environment, the sights, sounds, and smells grapple with my senses, and I am hit with a visceral gut reaction: I either like the space, hate it, or become more or less ambivalent to it. In that initial response, I ground my appreciation of and interaction with the space as a product of design.[7] After this first impression, when I engage in whatever activity the space offers, my time in the space is used to either confirm or reject my initial response.

Moreover, as we move through designed spaces, the actions we take are often communicated to us by the space itself. In the waterfall example above, my student parked in a makeshift gravel parking lot on the side of a small mountain road. She began her journey at a wood-carved sign at the trailhead. She followed a slightly worn trail using only minimal markers and no map. She gazed upon the waterfall from a particular vantage point in a clearing at the bottom of the falls. The trail's design controlled, or at least suggested and reinforced, her behaviors.

This observation that space impacts behavior is no secret to those who design spaces. The spaces of our daily lives constantly communicate a variety of messages with us, prompting us to behave in certain proscribed ways. Danish author and branding expert Martin Lindstrom details these messages in the context of retail shopping. He suggests that companies use all sorts of psychological and scientific tricks to persuade shoppers to buy things they do not need. These include things like signage, deals, and coupons, but also modifications of space. For example, changes in flooring, from smooth tiles to grooved tiles can cause a shopper to slow her cart instinctively. Designers refer to these areas as speed bumps. Speed bumps in a neighborhood cause drivers to slow down and pay attention. These flooring changes in grocery stores serve the same function, allowing the shopper to focus on a display or deal nearby. In one example, Lindstrom reported that using a grooved flooring speed bump paired with a sale display could cause a 45-second slowdown and result in an increase of up to 73% in spending habits.[8] I share this example to bring the conversation to the practical level of daily life. When teaching classes about experience design, I routinely take students on field trips to Walmart and Ikea to compare and contrast the user-experience design principles at play in different retailers. The result of these observation trips is often the same: the realization that space is designed to elicit particular responses.[9] In this case the intended response is consumer spending.

Several schools of thought differ in their characterizations of the relationship between space, user, and purpose. In their book *Educating By Design*,[10] educational researchers Carney Strange and James Banning illustrate three common design principles that define the relationship between built spaces and visitors. These three principles—architectural determinism, architectural possibilism, and architectural probabilism—exist on a spectrum of behavioral influence between actor and environment.

On one end of the spectrum rests architectural determinism. The viewpoint of architectural determinism suggests that the built environment is directly linked to the behavior that occurs in it. The choices we have in a built space—where to sit, how to move, how to interact with co-present others—are dictated by the physical structure. Adherents to this philosophy might assume that actions inside spaces are prescribed by the space. For example, when visiting your neighborhood pool, the security gate bears an element of determinism. Everyone coming in or out must pass through it. The pool area has no other means of entry. This design philosophy assumes that the environment is active and the user passive. The benefits of this concept for the designer are many. Among them: the space will be used as intended; human behavior can be controlled by design; and the designer has deterministic power. However, the difficulty with this viewpoint is the extreme passivity it requires of actors in the environment.

On the other end of the spectrum lies architectural possibilism. Possibilism suggests that a built space sets limits on the actions a user might take, but that the user is free to use the space in any number of unrestricted ways.

This design philosophy assumes that the environment is passive and the user active. Entries to the neighborhood pool through the security gate could come in any number of ways. The gate could be opened with a key card, hopped by an agile teen, or flattened by a steamroller. An entrant might dig a hole under the fence, parachute in from above, or pick the lock. These are certainly all possibilities for entry, pointing to the benefits of this design philosophy: The environment remains malleable based on the whims of actors; the environment accepts multiple viewpoints without restriction; and the actors control their own behaviors. But the difficulty with this viewpoint is the extreme passivity it requires of the environment.

The resulting middle ground is architectural probabilism. Probabilism assumes that the design of a built environment can increase the likelihood of some actions over others. In the example of the pool, entering through the security gate or hopping the fence are far more probable than parachuting into the pool or using a steamroller to knock down the gate. Using this philosophy, the security gate does not dictate or restrict actions, but it does make some behaviors more likely than others, depending on its design. If designed well, the users might interact with it as intended. Flaws in the design or its material construction might result in the gate being propped open by frustrated users or the creation of a secondary means of entry or egress. This probabilistic design philosophy assumes that the actors and environment are both active together—perhaps even interactive.

Strange and Banning point out that all three of these positions can provide insights into the interactions between environment and user. In addition, all three point to the role of design in articulating these interactions. A designer's belief about the behavioral influence of spaces on users is foundational to the designed outcomes of a space. Some of us may not like to think that our behaviors are being subconsciously modified by corporations, urban centers, home designers, and church architects. Nevertheless, most of us would agree that spaces we encounter are designed for particular purposes. Most spaces elicit some particular actions. I would argue that we even let spaces dictate our behaviors more often than we care to admit. If we give this much power to the physical dimension of space, the fact that we allow digital technology to dictate our movements should come as no surprise.

This book aims to explore the ways that digital technology shapes our choices about the use of space. These impacts are evident in nonverbal uses of physical space not only through the choices we make in designed spaces, but also through the movement of our bodies, our negotiations of space in interpersonal contexts, our choices about wayfinding and navigation, and in our connections to places. In each of these areas, we aspire to better understand the role of space and its uses—and the tactics we use to negotiate space in an increasingly digital world.

Computer Culture and Communication

Like spaces, digital technologies are designed with specific purposes to engage us in particular experiences. They act on us in ways that are deterministic, possibilistic, and probabilistic. They influence our cultural understandings of space and cause us to modify our use of space. These new experiences of technologies in space shape the ways we interact with spaces and others in them, and they can also be used to devise new types of spaces for us to encounter. Because we allow the design of spaces to direct, inform, and control our behaviors, emotions, and experiences in spaces, we likely translate that same affordance from the design of spaces to the design of digital technologies.

When we passively defer to the directions and whims of digital technologies, we transition our fatalistic acceptance of space into a digital environment. This deference is not all bad. We will relinquish our choices in favor of a pleasurable experience. For example, we might ask a global positioning system (GPS) to tell us the fastest route to a location. Having followed the turn-by-turn directions emanating from the device, we arrive at our destination easily and ten minutes ahead of schedule. But at what cost? Our knowledge of roadways? Spatial reasoning? Attention to landmarks? These have yet to be determined. But, technology, and our deference to it, offers benefits as well as consequences.

Research on the influence of technology on the human experience allows us to pause for consideration of who we are and how technology acts in our lives. MIT researcher Sherry Turkle describes her own reflections on computer culture this way: "My colleagues often objected [to my concerns], insisting that computers were 'just tools.' But I was certain that the 'just' in that sentence was deceiving. We are shaped by our tools. And now, the computer, a machine on the border of becoming a mind, was changing and shaping us."[11] Designers of digital technology are concerned with what a technology can do. Researchers of computer culture and communication are concerned with the impacts of designed technology on the human experience.

But academics like Turkle are not the only people pausing to consider the role of technology in their lives. With a wide variety of viewpoints, writers across genres, filmmakers, and popular authors are all contributing to the conversation, asking why we might choose digital interactions over physical ones. In *Real Simple* magazine's online site, writer Casey Stenger describes her own reflections from the decision about buying a phone for her 10-year-old daughter: "What am I giving my child here? It's not just a phone.... I am giving her a crutch for learning, a distraction from living, a hindrance to true exploration, a wet blanket on curiosity and worst of all a blockade to all, let's be honest, real human interaction."[12] Her own experience with being distracted by technology has led her to be concerned for the future. Meanwhile, Spike Jonze's Oscar-nominated film *Her* explores the possible future of emotional attachment to machines,[13] while the Supreme Court takes on cases surrounding the legal ramifications of the ubiquity of mobile phones as private data collection tools.[14]

The impacts of technology on our experiences are revealed through reflections—moments of pause that allow us to consider our actions and weigh the benefits or outcomes that emerge from them. Reflections, in turn, breed future actions that replicate or transform the actions that preceded them. Although some readers may see it as such, this study neither offers strategies for controlling visitor behaviors in spaces nor calls us to replicate practices that enhance the experience of users. Instead, it posits that our ability to reflect on the role of digital technology in our physical movements has the potential to change the ways we use technology by making our use of space more intentional, more thoughtful, and more purposeful.

As digital technologies become smarter, their roles in our lives become increasingly controversial. In an article for *The Nation*, Samir Chopra investigated the "personhood" of computer programs. He writes: "We interact with the world through programs all the time. Very often, programs we use interact with a network of unseen programs so that their ends and ours may be realized. We are acted on by programs all the time. These entities are not just agents in the sense of being able to take actions; they are also agents in the representative sense, taking autonomous actions on our behalf."[15] Chopra contends that computers not only aid us by taking actions we dictate, but also dictate our actions in an attempt to aid us. The search algorithms for online search, social networking, and online commerce sites are a good example of computers as seemingly benevolent support systems.

On Amazon, for instance, when I search for a book, the website's search function returns the book I searched for alongside a list of books others purchased when they searched for this book; a list of books I "might also like"; and a list of books in my search history. The algorithm pairs books for maximum readership based on my viewing and purchase history and its computation of my preferences. This algorithm worked well for me when I published a workbook to accompany a textbook I used in a strategic communication planning course. After one semester of assigning the two books together, students who used Amazon to purchase the same textbook at other universities saw that "people who purchased this book also purchased" the workbook I wrote. The algorithm increased my workbook sales, and I hope the book continues to help these lucky algorithm users in their work.

Likewise, Google's algorithm functions as a user-based search engine, giving different results to different users based on search history, location, and popularity. Digital organizer and author Eli Pariser famously described this in *The Filter Bubble: What the Internet Is Hiding from You*, in which he chronicled research about how the Internet privileges and hides various types of content from the average user.[16] Pandora Radio, a popular personalized radio service functions similarly by capturing user preferences through thumbs up (I like this song), thumbs down (I don't want to hear this song), and station selection. When I listen to "My Stupid Mouth by John Mayer" station, Pandora learns about my preferences by prompting me to listen to songs by artists I like, as well as similarly crafted unknown songs by a variety of other artists. Several artists have benefitted from this algorithm as I and other users purchase compelling records after hearing them on personalized streaming radio.

Social networking has also adopted user-based algorithms. Notably, on Facebook, a user's wall is filled not with the most recent posts by all of her friends, but rather by the most "important" posts determined by the site. Even though Facebook has yet to reveal its full algorithm, the site confirmed that factors like "how often you interact with the friend"; "the number of likes, shares and comments a post receives from the world at large and from your friends in particular"; and "how much you have interacted with this type of post in the past" among thousands of other measures contribute to the ordering of posts on a news feed.[17] For me, this algorithm blocks the posts of those annoying users who always rant about the latest ultra-radical political grievance, my acquaintance-only friends who I once knew in third grade, and the handful of individuals I blocked because I don't wish to know about their lives but need a vehicle for contacting them when needed.

These types of algorithms are beneficial to users, and they make our interactions in the digital world more pleasurable. But they also come with a cost. On Facebook, this cost was revealed when news broke that researchers affiliated with the site experimented on 700,000 users in 2012. The "creepy psychological study"[18] involved changing the user algorithm to manipulate user moods. Published in the *Proceedings of the National Academy of Sciences*, the study illustrates the role of "emotional contagion" on social networking, suggesting that emotional states can be transferred from user to user.[19] This study only added fuel to the continuing debate over digital versus personal control of information that is spreading across digital tools. On Amazon, authors and purchasers took issue with these algorithms when the book-selling giant allegedly de-privileged titles published by Hachette Book Group in site sales in a dispute over e-book revenues.[20] Purchasers were unable to pre-order books on Amazon by popular Hachette authors during the dispute. For Google search, Eli Pariser[21] admonishes users against blind reliance on the search algorithm, noting that while the site returns helpful results, the results may be skewed in favor of the searcher's personal preferences. The result would be that opposing information or viewpoints might be hidden from the user. The learning lesson for us all in these debates is this: the writer of digital code establishes the rules of user access to information.

Hence, the benevolent support system created by digital algorithms signals a shift in access to information. On one hand, the algorithms assist in developing usability and responsiveness, two criteria highly valued in digital tools. On the other hand, the algorithms control the means whereby information is navigated, privileging some information over others. Thus, the algorithm becomes the new gatekeeper for information in a digital society.[22]

For our purposes studying spaces, the question for us to consider is how digital interactions shape the ways we move. If digital codes can impact the way we navigate information, then they might also change the way we navigate space. If digital codes privilege certain information, then they might also privilege certain behaviors in space. If digital codes can suggest information we might like, then they might also suggest spatial interactions we might like or connect us to people and things in a space based on our preferences. In this genre of work, Stanford University researcher Bill Fogg, author of *Persuasive Technology: Using Computers to Change What We Think and Do*,[23] articulates the behavioral response of the user to technology as a term he coined: "captology." An acronym for *computers as persuasive technologies (CAPT)*, captology is interested in the ways that digital technology can change the attitudes, values, and behaviors of users. His Persuasive Technologies Lab at Stanford investigates what he calls behavior design—the concept that people can be persuaded by technology to behave in specific ways. His recent work includes research into mobile devices and the impact of captology for health benefits. This study aligns with Fogg's work in that it considers digital technologies to be persuasive technologies, shaping the ways that we interact with spaces we inhabit. However, this study is much more concerned with the variety of roles that digital technologies play in those spaces.

The benefits and consequences of embedded technologies are the scenarios that drove me to investigate the intersections of physical space, digital technology, and human experience. These intersections are formed in a fusion of spatial problem solving, human interaction design, aesthetics, digital manipulation, and ethical quandary. And, they play themselves out in diverse contexts and situations.

Rethinking Physical Space and Digital Technology

As the physical world and the digital world become increasingly intertwined, our use of physical spaces is ready to be studied anew. In the social sciences, the study of human behavior has taken a spatial turn over the last few decades. Hubbard and Kitchin describe this shift in the introduction to their compendium, Key Thinkers in Space and Place, as an "acknowledgement of the centrality of space in social theory" and a "broad engagement with social theory by geographers."[24] In this type of study, the fields of sociology, cultural studies, and geography collaborate and connect in the articulation of space and the meanings of people's use of space. Other traditional disciplines including literary studies, education, political science, and communication have much to add to the transdisciplinary discussion of space and place, and many researchers and thinkers are considering the ways that humans and spaces co-construct meaning.

This intersection between space and social theory provides multiple opportunities to consider the means whereby spaces and people operate together, and the role that digital technology assumes in shaping that interaction. Three themes emerge throughout this study and demonstrate their influence across the contexts presented herein.

First, the mutual impacts of space on us and us on space. The physical space we inhabit and our use of it cannot easily be separated. We understand spaces by being in them and experiencing them. Spaces act on us through our entry into the space and our choices once we inhabit them. We impact space by our participation in space or our lack thereof. And, our use of space is negotiated among the people present and between the environment and people within it. We can know about spaces or places, but that knowledge always leads to the question, "Have you been there?" The "being there" is the key to the mutual impacts between us and the spaces we inhabit.

Second, the lens of experience design as a framework for analysis. This book posits that a consideration of the space-technology-user interaction in both established and emerging spaces is an area primed for study and reflection. As a scholar, I choose to analyze our lived experiences through the lens of information and user-experience design. For those unfamiliar with information design and user-experience design, an appendix at the end of this book acts as a sort of primer for this concept, introducing the concept of information design and tracing the trajectory of its development as a field toward user-experience design. Many examples herein will draw from information and user-experience design theories to frame this analysis of communication in the context of space. As such, I hope to show that user-experience design offers us a new lens for the study of spaces as designed objects, as well as illuminates the relationship between designed spaces and the behaviors that occur within them. The language of experience design opens an engaging framework for studying spaces in myriad contexts and evaluating the role digital technology plays in the interactions between users and spaces. As we engage in this design thinking, pulling it back into the conversation throughout the text, the design of the interactions among humans, technologies, and spaces come into focus in some obvious and some surprising ways.

Third, the interaction between digital and physical. As digital technology becomes increasingly ubiquitous, the ways we operate in the physical world become more reliant on the digital. It seems almost silly to pose the two worlds as separate—a digital world and a physical world. For most of our experience, we interact with both, unless we make a conscious effort to unplug. The two worlds are continuously merging, but from time to time we experience glimpses of the physical world without the digital or the digital without the physical. This interaction might be described as a dialectic with the digital on one end and the physical on the other with our lives negotiating a balance between the two. Or, perhaps it might be described as two intertwined sources of information and knowledge. Luddites might even describe it as pollution of purity in the physical world. Futurists might see the interaction as the inevitable progression toward singularity. However it is viewed, the interaction between digital and physical is one that shapes our lived experience. At times, this study will take each of these and other approaches to understanding the merger of physical and digital, but the concept remains vital to an understanding of digital technology's impact on our lived experience.

Our relationships with spaces, an understanding of experience design, and the merger of physical and digital provide a vibrant intersection of theory and practice in the study of space. The chapters in this book are situated in the spatial and proxemic contexts formed by this intersection. I invite you to consider these contexts with me, to evaluate your own experiences in the spaces you inhabit, and to explore how digital technology shapes the ways we move.

· 2 ·

DIGITIZING PROXEMICS

Dismayed by a series of negative online reviews citing long wait times, an anonymous restaurant manager[1] in East Midtown Manhattan searched the restaurant archives for answers. After investigating receipts, reviews, and feedback from staff, the restaurateur[2] and a consulting firm turned to the restaurant's cameras. They compared two sets of surveillance footage, one from a dinner service in 2004 and one from a dinner service in 2014, in an attempt to discern the cause of bloated wait times for patrons. The observed differences in dinner service times in 2004 versus the same service in 2014 were striking. Over the decade, the average time for food preparation remained constant. Yet, over that same decade, the average time a customer sat before placing an order increased by almost 200%, from 8 to 21 minutes. And, the average time customers spent with food on the table increased by 20 minutes. These changes resulted in an almost doubling of the average total mealtime observed in this restaurant, from 1 hour and 5 minutes in 2004 to 1 hour and 55 minutes in 2014.[3]

Based on the content of the surveillance footage, the investigators found one noteworthy difference in the customers of ten years ago and the customers of today: the presence of mobile phones.

When asked about the way mobile phones have changed the restaurant business, I might typically reference the ability to make reservations, search menus, and read customer reviews, or perhaps the ability of restaurants to market through location-based and user-targeted channels. These all seem like general benefits for restaurant owners. After all, Internet communication has given restaurants (and all businesses) direct business-to-consumer lines of communication. It has allowed restaurants to market directly to their clients and to access volumes of information about their target audience. In addition, social and participatory media have allowed satisfied clients to become marketers for their favorite eateries. But, from an experience design perspective, the impact of mobile phones on in-restaurant dining represents a dramatic shift in the way patrons use their time, how they orient their bodies, and even the speed at which they eat their meals.

In the East Midtown restaurant's surveillance footage, customers could be seen spending several minutes at the beginning of the dinner service connecting to the restaurant's wireless Internet and asking the wait staff for help. The wait staff reported spending more time explaining wireless login procedures than they did explaining menu items. Customers looked up ingredients on Google before ordering. Waiters snapped photos for patrons, and patrons took selfies, photos of their food (the early foundation of Instagram), and images of each other. In fourteen instances in the footage from 2014, extended photo shoots resulted in the food being sent back to the kitchen for being "too cold." Thus, the average dinner service time increased. And, in 2014, as they were exiting, eight patrons even bumped into waiters or other patrons while simultaneously walking and operating their phones.

The anonymous restaurateur summed up the experience with a plea to customers on Craigslist: "Can you please be a bit more considerate?" The plea belies a sense of angst associated with technology's invasion of our lived experience in the spaces we inhabit.

In 1966, Edward T. Hall coined the term "proxemics" in his book, *The Hidden Dimension*.[4] Hall was a cultural anthropologist. His work focused on space as a cultural marker of learned behavior. For Hall, proxemics was the study of the nonverbal use of space as a marker of culture. He observed interactions occurring in spaces to ascertain the way that use of space was negotiated. Hall read and shared research about the differences in perspectives on land versus inside submarines. His analysis included pigs feeding together at a trough and people sitting together at a table—studies not unlike the self-study conducted in the East Midtown Manhattan restaurant. And, he classified the variety of information present from other nonverbal cues like eye contact, body heat, and time as they relate to proxemics.

As a result of the widespread acceptance of his work, proxemics is now a widely used term in the canon of communication studies and sociology, and is often referenced alongside nine other forms of nonverbal expression. A list of nonverbal forms of communication typically includes[5]:

- kinesics (body language including facial expression)
- proxemics (the use of space),
- olfactics (the use of smell),
- haptics (the use of touch),
- chronemics (the use of time),
- artifacts (items of significance),
- physical appearance (generally related to cleanliness, dress, and demeanor),
- situational factors (light, sounds, temperature),
- paralanguage (utterances that are not words, or sounds that shape words like accents, tone, and pitch), and
- silence.

These ten nonverbal forms of communication describe the ways that human communication operates alongside and beyond verbal messages, the words we speak. Our nonverbal messages often say far more than the verbal messages we choose to share. Nonverbal communication is often discussed in training experiences for people across a wide variety of disciplines. In public relations courses, textbooks usually devote attention to cultural differences that might impact global public relations and the importance of learning cultural norms for body orientation, eye contact, and touch. Organizations and human resources departments offer trainings in proxemics to promote interpersonal and intercultural understanding.[6] In health communication, researchers have compiled resources for physicians to encourage the successful use of nonverbal expression to improve the patient-physician interaction.[7] Even popular culture attempts to educate us on the use of nonverbals for personal gain. The film *Hitch*[8] opens with a voiceover sequence in which the title character, a matchmaker, informs us that "60% of all human communication is nonverbal. Body language. Thirty percent is your tone. So that means that 90% of what you're saying ain't coming out of your mouth." For the rest of the movie,

Hitch spends his time instructing would-be suitors in the art of nonverbal expression. These are just a few examples of the more or less practical applications of nonverbal communication to everyday life.

Like all nonverbal messages, proxemics interact with the other nonverbal and verbal messages present in a situation. For example, when we use sarcasm, our nonverbal messages contradict our verbal message. Words said with sarcasm rely on nonverbal messages of paralanguage (tone) and kinesics (facial expression) to illustrate the contradiction between the actual words spoken and the intended meaning. These interactions make the study of communication extremely complex. For this reason, nonverbal communication scholars will often try to isolate the effects of one or more of these nonverbals to try to understand how each might function in a given situation.

If we conducted a critical analysis of the East Midtown Manhattan restaurant study mentioned above as the restaurateurs did, we might focus on the nonverbal forms of kinesics, proxemics, and artifacts and the ways they converge to influence chronemics. On the surveillance footage, kinesics might include the body language and posture of the patrons and waiters. In addition, kinesics would encompass emotional displays of facial expressions including happiness, sadness, anger, and boredom. These messages give clues about the types of verbal interactions that might have occurred at the tables, between patrons and servers, and among the staff. Proxemics incorporates the space of the restaurant and dinner table as well as the negotiation of distances between people. As examples, patrons might huddle together for a photo, sit on the same side of a table or booth, or reorient themselves for the different purposes of using mobile devices, shooting photographs, conversing, and eating. The distances between waiters and patrons might also inform analyses. Artifacts in the situation, namely mobile devices in this case, shed light onto important practices and behaviors. Their absence in 2004 and presence in 2014 indicates an interesting correlation—if not also causation—between artifact and behavior. These three things converge to influence the timing of dinner service. If we moved the study one step further into applied research or strategic communication planning, we might not stop alongside the Craigslist poster but rather suggest a modification in one form of nonverbal expression to cause changes in the others. For example, could a change in proxemics alleviate some of the grievances concerning the artifacts in question? Perhaps so.

Even though proxemics is a widely used term among scholars of nonverbal communication, sociology, and other related fields, I recognize that its appearance often sends people to find a dictionary, and usually on their mobile phone. For our purposes herein, proxemics will be defined as the study of space and the use of space in human interaction.

Edward T. Hall's Proxemics

In *The Hidden Dimension*,[9] Hall grounded the work of defining and describing proxemics in relation to his observations and those of others interested in space. In his work, he began a conversation about the seemingly hidden role of space in our lived experience. Among many astute observations he made about space, some of his most enduring concepts include classifications of the human uses of distance and the organization of spaces as territories.

Human Distances

Hall's theory surrounding the distances of man was based on observations of people interacting together. He charted the relationships between how people stood in space and field of vision, thermal awareness, volume of speech, ability to touch, linguistics, and other characteristics of interpersonal interaction. The result of these relationships became Hall's classification of 4 major categories of space: intimate, personal, social, and public. These categories largely shape our understanding of proxemics in the years since.

Intimate space is the 18 inches closest to one's body. Two people interacting in intimate distance are embracing, whispering, and likely touching. At this distance, one person can feel the body heat of another and smell subtle fragrances. Two people face-to-face at intimate distance can see the details in one portion of the face, for instance the markings of the pupils of the eye, and can even get close enough for vision to be unfocused and distorted. They can communicate in whispers and breaths. Hall characterized the 6 inches closest to the body as close intimate space, opposed to far intimate space in the 6–18 inches range.

Beyond intimate distance, personal distance is roughly any space in arm's reach. This area, 1.5 to 4 feet from the body, is generally reserved for conversations and interactions with close family members and friends. In contrast to intimate space, in the range of personal space, two people standing face-to-face could discern each other's full faces at once. They could verbally communicate in a casual voice or a quiet conversation voice. Hall divided personal space into close personal space—think elbowroom—and far personal space—within arm's reach.

The boundaries of intimate and personal space might best be defined by our reactions when invasions or violations of these boundaries occur. We readily recognize invasions of these spaces, and react. This reaction might be a polite, "Excuse me," a counter-movement, or any number of verbal or nonverbal responses. Moreover, the closer another person moves in relation to our bodies, the more emotional connection we feel—for better or worse. Invasions of personal space by a romantic partner might be welcome, and the anxious energy of the situation would be heightened at intimate distance. Likewise, invasions of personal space by an angry friend would be threatening, but angry invasions of intimate space heighten the severity of the threat.

The boundary between personal and intimate, Hall suggests, is culturally bound and reflective of our learned experiences in society. He points out that American children often invade each other's personal space during play whereas American adults react unfavorably to feeling cramped on a bus or subway. However, members of some other cultures may not share Americans' displeasure with closeness, while still other cultures would prefer greater separations of distance.

Beyond personal distance, the space 4 feet to 12 feet from the body is described as social distance by Hall. We might expect interactions among acquaintances, colleagues at work, and group meetings to take place at this distance. At social distance, a group of people can generally carry on a conversation without raised voices. Each person can see multiple faces in the field of vision and focus on each other person. Hall again divided social space into close social space, characterized as 4 to 7 feet, and far social space, from 7 to 12 feet.

Outside of the 12 feet closest to the body is public space. Generally speaking, we allow people to move and operate at public distance without much concern. Within 25 feet, which Hall called close public space, a loud, full speaking voice is required to get another person's attention. A person's entire body can be discerned at once. Beyond 25 feet, far public space, while we can discern that people are there, we might not be able to distinguish unknown individuals or specific voices. In this far public distance, a speaker would need to exaggerate gestures and amplify sounds to be seen and heard. Typically, practical uses of far public distance for interaction would be reserved for public performance.

The boundary between public distance and social distance is revealed when we internally assess the intent of someone's presence in our lives. Crossing this boundary triggers a subconscious response in which we quickly consider whether to greet, fight, or run. In most scenarios, when people cross from public space into social space, we might offer a greeting or smile, or perhaps indicate nonverbal avoidance depending on the social and cultural context. Southerners, those people born and raised in the US Deep South, who visit New York City for the first time always remark about the cultural difference they note. The story usually goes something like this: "When I walk around the streets in South Carolina, everyone smiles and says hello, but in New York City, no one even made eye contact with me. I thought if I said hello, someone might shoot me!"[10] The difference my friend noted here is one of culturally defined proxemic interaction. When a person moves from public space to social space, what happens? On the streets of South Carolina, the movement across the boundary from public space to social space may be actively welcomed with a smile and eye contact if not also a warm greeting. However, on the streets of New York City, the crossing of the public-to-social-space boundary may be actively ignored.

The classification of these proxemic distances is foundational to an understanding of Hall's proxemics as a whole. Moreover, these distances are some of the most widely referenced ideas in Hall's work on space.

Organization of Spaces as Territories

Hall argued that not only is the study of proxemics interested in the ways that people negotiate interpersonal space and distance, but also in the organizing principles they use in arranging their spaces at home and work, and even in the organization of their cities and towns. He was particularly concerned with territory and territoriality. The word "territory" evokes ownership and power, and Hall noted that territories could include a variety of boundaries and designs.

When thinking about territory, Hall categorized space into three genres of design: fixed feature space, semi-fixed feature space, and informal (or dynamic) feature space. Fixed feature space is typically characterized by immovable design. Semi-fixed feature space offers furnishings and boundaries that can be moved with varying degrees of effort. Informal space is often absent of fixed or semi-fixed features, or the territory is dynamically fluid, allowing it to change purposes for different people at different times.

Fixed features of space can be real or perceived, tangible or imagined. The exterior wall of my house is a fixed feature. It is generally immovable but also quite tangible. It sets a real boundary between inside and outside. Likewise, the property line that runs between my yard and that of my neighbor is also a fixed feature. The property line is immovable even though it is intangible. But it is just as real as the wall. Fixed features can be placed by acts of construction, demolition, law, culture, geography, and design. Their commonality is that they are perceived as immovable.

Semi-fixed features are moderately movable. Typically, furnishings within a space fall into this category. Placed into different arrangements, semi-fixed features balance the aesthetic of a space with its function. To further our understanding of semi-fixed feature space, Hall pointed to social arrangements of space, or sociospatial design.[11] Sociospatial design considers the effects of semi-fixed feature arrangement on human interaction, classifying arrangements on a spectrum from sociopetal to sociofugal.

In a sociopetal arrangement, people are oriented toward each other for social interaction. Conversely, a sociofugal arrangement of space orients people away from social interaction. Imagine four chairs in a den or living room. If these chairs were arranged in a circle-like formation, people could sit on each chair, see all of the other people in the seats, and hold a friendly conversation. This would be a sociopetal arrangement because the chairs orient people toward one another. However, if the chairs were positioned facing each corner of the room, they would be in a sociofugal arrangement, orienting people away from each other. Perhaps a host might try this arrangement if all the guests are three-year-olds who each need a time-out.

The common misconception of sociospatial design is that sociopetal arrangements are "good" and sociofugal arrangements are "bad." This is simply not the case. Sociofugal and sociopetal arrangements, in balance, provide a variety of important opportunities for interaction. Both are necessary in balance to orient people in a space. I tried a little everyday experiment of sociofugal and sociopetal space in my town. Townes Plaza in downtown Greenville, SC is named for Charles Townes, the inventor of the laser and resident of the city. The grassy plaza is bordered on one side with benches. Also on the plaza is a Starbucks Coffee with outdoor seating. The benches and the outdoor seating occupy the same physical location, meaning that people sitting on the benches and at the outdoor seating are in close proximity to each other at a personal space distance. When the plaza was constructed, the benches were turned in a sociopetal arrangement facing the outdoor seating. The result was that someone could sit on the bench and be extremely close to and oriented toward a group of diners at a table. Thus, the benches and tables were seldom used simultaneously. As I am prone to do, I reoriented the (heavy) benches in a sociofugal arrangement so that people sitting on the benches faced away from people sitting at the tables. Within ten minutes, all of the benches and tables were full. While the city has never thanked me for this contribution to the user experience in Townes Plaza, the sociofugal arrangement in this area promotes more interaction with the plaza than the sociopetal arrangement—even though both occur inside personal distance. As you might note, sociospatial design blends elements of proxemics including location, distance, orientation, and movement.

Informal spaces are spaces open for interaction that contain dynamic features. Hall doesn't spend a lot of time discussing informal spaces, but rather suggests that human use of the distances previously discussed can be viewed unobstructed in informal spaces. In a later work,[12] Hall reclassified this third type as dynamic feature space. Dynamic feature spaces are those that are highly changeable, modifiable, and adaptable. Nomadic lifestyles rely on dynamic features and territories that can shift ownership. Spaces that are highly flexible might be considered dynamic when they shift and change to meet the purposes of the user.

Proxemics as Lived Experience

Together, these two classifications—human uses of distance and organization of spaces—serve as a foundation for proxemics. They offer a starting point for the study of space as a component of lived experience. Hall's applications for this study were many: linguistics, urban planning, interior design, communication, sociology, architecture, workplace training, intercultural dynamics, and the list continues. At the time of his writing, Hall's concepts of proxemics emerged as an inspiring moment in anthropology. Commentators responding to his appropriately titled article, "Proxemics," in *Current Anthropology* in 1968[13] heralded the concept. They described the naming of proxemics as the revealing of a component of human interaction that was previously hidden. Still today, many people who encounter instruction in proxemics in an introductory sociology or communication course recognize it as a hidden component of culture and relationships that, once revealed, helps articulate spatial interactions.

In *Cities for People*, architect Jan Gehl notes that the ground rules of proxemics are common in much of lived experience: "We use these common ground rules for communication in all of life's situations. They help us to initiate, develop, control, and conclude relationships with people we know and don't know, and they help us signal when we do or do not want contact.

The existence of these common ground rules is important in order for people to move securely and comfortably among strangers in public space."[14] Some of the most obvious times in life when we recognize the work of proxemics are awkward encounters with strangers in enclosed or defined spaces. For example, when I step onto a crowded elevator, I consider where to stand and how to orient my body. Proxemics is at play as I decide whether to face the doors, face another rider, or squeeze into a corner. If I approach a family-style seating arrangement in a busy restaurant, proxemics helps me decide where to sit. When I join a group of people in conversation, proxemics is the key to discerning how close I should stand to the other members of the group. Proxemics impact our decisions across spaces and punctuate our experiences with others.

Wrapped up in the concepts of proxemics are several concepts that appear throughout a discussion of the subject: location, orientation, movement, distance, navigation, wayfinding, and identity. As these terms will appear repeatedly in the conversation surrounding proxemics, let's define each in the context of the study of the human use of space.

Location refers to the specific point in space occupied by a person or object. When we define a location, it is usually described in one of two ways. First, location might be defined as geographic location, the specific point on a map generally defined by latitude and longitude. Location might also be defined as a specific point in space relative to other points. For example, if I was hoping to be found in a multi-story office building, I would need to supplement the geographic location with a description of my altitude: "I'm at One Main Street, on the 7th floor." In this way, geographic location is different from relative position, and both together can be used to describe location.

Orientation refers to the directionality of physical position. When we describe orientation, we typically reference the direction of our gaze relative to some fixed point. For example, to watch a sunrise, I would face east. In terms of semi-fixed feature space, sociospatial designs rely upon orientation. Remember, in sociopetal arrangements, I would face toward others in the space. In sociofugal arrangements, I would face away from others in a space.

Movement refers to any change in location, position, or orientation. When we describe movement, we often discuss the original position or location, the direction and velocity of change, and/or the final position or location.

The space between two fixed points, positions, or locations is referred to as distance. Distance between humans, in Hall's terms, is physical and could best be described in tangible terms by the sensations that could be experienced by two people standing together. Distance can also describe the space between humans and objects, multiple objects, or multiple locations.

Navigation is the joint act of, first, ascertaining one's current position and, second, moving along a route to get to another position relative to the first. On a cross-country journey, a driver might navigate his car from Charlotte to Seattle by taking day trips from point to point. An employee might navigate her way to the exit while dodging particular co-workers. Related to navigation is wayfinding. Wayfinding is a subset of navigation that describes the human processes of discerning a route in an environment.[15] Wayfinding encompasses the decision-making aspects of navigation and the signs and symbols that help us plan a path for navigation. Many design professionals think of wayfinding as the art of providing directional cues for users. For example, signs identifying entrances, exits, and restrooms would all be classified as wayfinding tools as would street signs, crosswalks, and maps. For our purposes, wayfinding will refer to directional cues and the resulting process for determining which way to go and navigation will refer to the act of moving from point to point.

Identity, in the context of proxemics, relates to the self and the ability to discriminate among multiple people, multiple locations, and multiple objects present in space. Each person, location, or object has a specific identity. As we discern the identities of the people and objects in a space, as well as the features of a space, patterns of behavior emerge and change relative to identity. Identity—the process of knowing and revealing components of self—shapes our interactions with other people in spaces and is, therefore, worthy of inclusion as a foundational concept in proxemics.

Location, orientation, movement, distance, navigation, wayfinding, and identity all shape the ways we move in, around, and through spaces. They shape the interactions that we have with others in space, and they articulate specific components of our use of space that are valuable to our study. These seven items are not meant to be an exhaustive list, but rather a starting point for some common language in this study.

Theorizing Digital Proxemics

In the half-century since Edward Hall coined the term, proxemics has remained a fairly static concept. However, as he notes, the characteristics of proxemic interactions are markers of culture. As culture changes, so change the rules of proxemics.

Cultural Shift

Over the last several decades, the communication and information culture has shifted rapidly. Perhaps that sentence is an understatement. In American culture, the 1980s and 1990s were eras of rapid influx of digital technology into every aspect of life. As an example of this dynamic shift in lifestyle, consider children born between 1976 and 1984. Most children in America born before 1976 (Generation X[16]) were in high school before personal computers made their debut in schools and homes. Most children in America born after 1984 (Millennials) never knew a school classroom before computers, and many had a computer, or multiple computers, in their homes. That almost-a-decade of births in the middle, children of late seventies and early eighties, grew up learning about personal computing technology alongside the designers of personal computers. We experimented on Apple 2e computers in primary school and played Oregon Trail on a green screen. Children born in this decade still remember how to use a library's now obsolete card catalog and were on the cutting edge of internet search algorithms. We existed in two worlds, one analog and one digital. We weren't digital natives, but we also weren't digital immigrants.[17] Maybe we were digital pioneers (which would give blogger Anna Garvey's moniker for this micro-generation—"The Oregon Trail Generation"[18]—a lot more street cred and a whole new meaning).

Technological changes in physical hardware demonstrate the shift from analog to digital in broad strokes. Cassette tapes became CDs. VHS tapes transitioned to DVDs.[19] CDs and DVDs transitioned to downloads. Phones connected by cords to the wall became wireless. Pay phones were the preferred method of communication from anywhere outside the home. Huge bag phones entered cars for a few years so we could talk on the go, followed by flip phones, and then by cell phones with retractable antennas, making them small enough to fit in a pocket. We used to find phones to place calls. Now phones find us to receive calls. These are just a few simple representative examples of the shift in personal technologies that marked the late 1980s and early 1990s. Still today, we sit in awe of the power of technology and reminisce over life before [insert technology here: mobile devices, social networking, broadband Internet, laptops, tablets]. Reminiscing about technological advancement is a marker of rapid cultural shift. Culture fundamentally shifted because communication technology fundamentally shifted.

But the shift toward digital and Internet communication is one in a series of communication technologies that caused dramatic shifts in human history.

Media theorist Marshall McLuhan chronicled these and predicted some of the technological shifts noted in the last few decades.[20] In his history of the epochs of man, he demonstrates how humans moved from tribal to literate to electronic. The invention of the alphabet transitioned humans from oral historians to written-word poets. The invention of the printing press in 1453 AD was the end of the Dark Ages and spawned the Enlightenment of the Western world. The invention of the telegraph in 1835 AD allowed humans to send information electronically, ushering in an era of broadcast technology from telephones to radio to television. Even though he wrote about these epochs in the 1960s, he theorized that a new age was on the horizon. Today, we often call this the digital age, following the widespread adoption of the Internet in 1994 AD which allowed information to be connected in networks rather than broadcast from source to viewer.

Each of these culture-changing inventions was decried in its time as the end of human civilization. In the late fifteenth century: *If everyone can read books, it will be an end to moral order!* In the early 20th century: *The radio and television will rot our brains! No one will read books anymore!* At the end of the 20th century: *The internet will disconnect us from people and ourselves!* Instigators of these messages were quite right, although arguably over-zealous. In some cases, like that of the East Midtown Manhattan restaurant, changing technology can be credited with both perceived benefits and perceived repercussions. Digital culture is still in its early days and thinkers around the world are debating what its true characteristics are. However, cultural theorists agree that, like other technologies,[21] communication technologies change cultures.

Digital Proxemics

This study enters into a discussion of the impacts of technology on culture through consideration of the marker of culture described by Edward Hall: proxemics. By pausing to examine the ways proxemics are being used in the digital age, we can reflect on the impacts of digital technology in our lives. Proxemics can be both a marker for and a microcosm of a changed culture in the digital age.

In the genre of digital culture and proxemics, current thinking emerges from a wide variety of fields and disciplines. As an example of this intersection, in the *Journal of Research Practice*, researcher Tiiu Poldma[22] describes interior spaces through design thinking, offering thoughts about the role of the user in the use and social context of the space. Presenting a case study about the interior design of a nursing home, Poldma concludes that attention to user needs and contexts can shape the design of a space and the interactions that occur in it. "In this technologically and digitally enhanced world, objects are transitory, spaces can be virtual or physical, while communication and interactions are varied and changing constantly, all affecting social and political norms... Spaces are no longer designed for one specific use, nor as the determinant of a particular set of activities."[23] Our experiences of space, she notes, involve overlapping tasks that transcend location and time, reality and imagination, and physical and virtual.[24] In short, Poldma argues that our experiences of the world and norms of our behaviors are dramatically impacted by digital technology.

As a second example of this conversation, in *Smart Mobs*, acclaimed futurist Howard Rheingold describes the transformation of culture that could accompany the rise of sentient devices—those inanimate objects imbued with thinking power. He writes, "These projects... are converging on the same boundary between artificial and natural worlds," and described avenues for possible research on items which span the physical-digital boundary: (1) information in places; (2) smart rooms; (3) digital cities; (4) sentient objects; (5) tangible bits; and (6) wearable computers.[25] Each of these items is worthy of much more conversation, and we will return to them later in various contexts. For now, these six vectors represent invasions of the physical world by digital technology. And, note that Rheingold was writing about these projects in the early 2000s, before most of us had reliable service on our extra-large flip phones. Each vector considers one possible means whereby artificial technology can become an actor in the human world.

As the two examples above demonstrate, digital proxemics is not an unknown area of study, but it is relatively undefined, and reflection on the role of digital technology in our spatial interactions is extremely limited. Many good reasons exist for this lack of reflection. Most notably, research thus far focuses on the ways that digital technologies can create experiences in spaces. Researchers and designers have been more interested in determining what digital technology can do *for* us and less interested in what the same technologies do *to* us.

For example, the fields of pervasive computing, human-computer interaction, and ubiquitous computing, or ubicomp, aim to understand the integration of technology into daily lived experience and to design new possible interactions between humans and smart things. In recent years, the trajectory of the field was chronicled by a group of researchers at the University of Oulu in Finland and presented at an ubicomp conference in Seattle, Washington.[26] Their analysis indicated that the field was historically focused on location-aware applications, network challenges, and user contexts, with recent shifts toward sensing, crowd sensing, and data analysis. Ubicomp designers are creating technologies that have the potential to change the ways that we interact with the Internet, the physical world around us, and other people.

In the popular marketplace and the growing Internet of Things, wearable technologies are on the cusp of rampant adoption. Sentient objects like smart thermostats, adaptive windows, and responsive vending machines are entering our homes, workplaces, and restaurants. These technologies intersect with education, art, business, healthcare, and every other facet of daily life. And these are all good things.

My concern, and the concern of this book, is that we might cede our control in the spaces we inhabit to digital technologies without realizing it. When we visit a museum, watch a film, or read a book, we are typically setting aside time for exploration and experience. Likewise, technologies in the spaces of life are designed to give us particular experiences. But we may simply move with and through these experiences at the whim of an unknown designer. In our daily lives, I worry that we may be experiencing the benefits of digital technologies in the moment without reflection.

A theory of digital proxemics reveals the hidden dimension of our use of space in our merging physical and digital worlds. To do so, digital proxemics asks the question, what is the impact of digital technologies on the relationship between user and space?

As I collected, compiled, and conducted research for this book, four categories emerged describing the possible interactions between digital technology and humans in the context of proxemics: informing, capturing, altering, and controlling.

First, digital technology embedded in a space can **inform** a user by converting data about the space into useful information. A space reveals information to a user through multiple means including wayfinding cues, fixed and semi-fixed features, and its unique characteristics. This information might include directions which orient a user to a space, narratives which provide information about objects in a space, or data about the space and its uses offered to the user for analysis. The act of informing revolves around the routes of information sharing in a space. As each of these areas becomes imbued with digital technology, the availability of information about the space and the speed of its delivery increases.

Second, digital technology embedded in a space can **capture** data about a user in the space and convert it into information. The act of capturing encompasses the role of technology in data acquisition from the user. Technology-based capturing might include motion and heat sensing, requesting data entry from the user, and a variety of methods of video surveillance. Mobile devices can provide opportunities for spaces to harness not only geo-location but also a wide variety of data stored online concerning user preferences and histories.

Third, digital technology embedded in a space can **alter** the behaviors of the user in the space. The act of altering signifies a change in the user's interaction with space. The concept that spaces shape user behavior has been clarified here already. Spaces were influencing behaviors for centuries prior to the arrival of digital technology. Thus, this category is specifically interested in the ways that digital technologies alter the space-user interaction.

Fourth, digital technology embedded in a space can **control** a user's interaction with the space. The act of controlling suggests that the emotions and behaviors of users in a space can be directed by digital technology. In the case of immersive virtual reality spaces, technology has full control, and users have full knowledge that technology is in control. However, in spaces that are hybrids of digital and physical space, users may be less mindful of the ways they are controlled by embedded technology. This fourth category is, by some estimates, the most controversial concept herein. It raises ethical questions about not only how we use technology but also whether or not we should. Thereby, it warrants special attention in each context to follow.

As I was studying these four actions—informing, capturing, altering, controlling—the converse of all of these effects became abundantly clear. Not only can digital technology in a space inform, capture, alter, and control the user, but digital technology accessed by the user can inform, capture, alter, and control the space as well. Spaces are embedded with digital technologies that inform, capture, alter and control the conditions of the user and thus influence the user's experience in the space. People carry digital technologies that inform, capture, alter, and control the conditions of the space and thus influence the user's experience in the space. Digital technology present in spaces and on people shape the interaction between users and spaces.

Therefore, digital proxemics might be defined as a digitally inscribed negotiation of personal space through the mutual mediation of actions between space and user. More simply stated, digital proxemics studies the impacts of digital technology on humans' experiences of space and its uses. As such, digital proxemics provides us a valuable starting point for reflection about this vibrant intersection of space, technology, and user experience.

Remember that Edward Hall suggests proxemics as a hidden dimension of communication. The hidden nature of proxemics requires observation and reflection to call it to mind. This reflection reminds us of the importance of space as a marker of our culture and lived experience. As digital technology continues to mediate our lived experience in space, bringing proxemics out of hiding and into our conversation becomes vitally important. Digital technology shapes the ways we move in, around, and through spaces. The ways that this shaping occurs are many and varied. The next several chapters present contexts which reveal the influence of digital technology on our understanding of proxemics. Our goal, then, will be to use these contexts surrounding the study of digital proxemics to reveal and reflect upon the multiple and changing roles of digital technology in our lives.

· 3 ·

DISTANCING OURSELVES

My first ride on a subway was on the Metro in Paris, France. We were travelling from our hotel to the Anvers station near Sacre Couer in the hilly Montmartre region of the city. Luckily for me, I was not in charge of navigating the Paris Metro on my own for my first subway experience. I remember two things about the experience. First, the view of the city from the hilltop basilica captivates the eyes in the early evening, just as the illumination of the Eiffel Tower begins. Second, the Metro doors seemed to close awfully fast. Several members of our group ran to board the train, but the doors closed and the train moved onward. As we left them behind, in my naiveté surrounding Metro transportation, I guess I didn't realize that they could just catch the next one.

Since then, I've enjoyed a healthy fascination with rapid transit systems, their navigation principles, the designs of their maps, and the role of rapid transit in urban centers. But none of these interests can top the unique experiences that rapid transit offers in the negotiation of interpersonal distances. Take a ride on the subway, underground, metro, or train in an urban center and you can practice this negotiation and experiment with its principles.

The subway car is a closed environment[1] in which we negotiate space each time people enter or exit. Then, we remain still while we ride. This alternation between stillness and movement is an added benefit for observers because the results of our interpersonal negotiation of distance hold their form for a few minutes as the train is in motion. When new variables are introduced at each stop, new spatial negotiations ensue. The doors close and we freeze in place again as the train begins to move, making rapid transit an excellent venue for proxemics observation.

When you observe carefully, you might notice the ways that people position themselves in the subway car. Riders tend to space themselves at regular intervals that offer relatively equal distances between people. In a car where only three people are riding, riders might position themselves throughout the car allowing other people to enter social distance boundaries without crossing the threshold of personal distance. When a fourth rider boards, she might move to the end of the car or space herself apart from the other three, giving each person relatively equal personal space to occupy. Later, when the car becomes standing room only, people still space themselves at (increasingly smaller) regular intervals. The distance between people shrinks from social distance and moves within the boundaries of personal distance. Note that this invasion of space is still negotiated by giving each person relatively equal space to occupy. You've probably noticed this phenomenon before. Birds congregating on telephone wires also exhibit this characteristic regular spacing behavior. While a ride in the full car might cause more discomfort than the ride in the empty car, interpersonal distance is still negotiated similarly in both situations through spatial arrangement.

A deeper look into the experience of the subway car allows the importance of other related factors to emerge in relation to the ways we arrange ourselves in a closed space. First, whether a nearby other is positioned seated or standing and their orientation both make a difference in the way I negotiate interpersonal distance and orient my body. For me, I would rather stand closer to someone standing than to someone sitting. If I have to stand close to someone seated on a subway, then the orientation of my body—how my body is turned in relation to their body—makes a huge difference for both of us. Second, position and orientation intersect with personal characteristics, such as height, weight, gender, and physical appearance, to impact our choices about interpersonal spacing in some obvious and some surprising ways. In deciding who to stand closest to, our internal and unspoken judgments, biases, and personal motivations might be physically observable. Nonverbal negotiation of interpersonal space offers a lot of information about how we, as humans, approach others of similarity and difference across a wide variety of physical characteristics. Third, layered on top of all of these factors, our knowledge of the others in the environment changes the patterns of orientation and spacing. For an example of all of these factors merging together, consider two males of similar build standing in the subway car. If the pair are strangers, they are more likely to stand back-to-back, face-to-back, or shoulder-to-shoulder than face-to face; whereas two male friends of similar build might be more likely to stand face-to-face or shoulder-to-shoulder rather than in face-to-back or back-to-back configurations. In addition, in a shoulder-to-shoulder stance, strangers might be more likely to align themselves in a straight line or angle away from one another whereas friends traveling together might prefer to align themselves angled toward one another in a facing formation. If this male dyad is replaced with another dyad of different characteristics based on either body position, height, body size, gender, physical appearance, or knowledge of each other, their exhibited proxemics behaviors will change. The crowded subway car is a tiny but fascinating microcosm illustrating the ways people negotiate interpersonal distances in daily life.

We'll discuss some of the ways this negotiation occurs and look to see how digital technology changes these motions. By applying the concepts of digital proxemics to the position and distancing of our bodies, we can see how digital technologies relate to our movements both within the spaces we inhabit or visit and among the others present in the space.

Negotiations of Interpersonal Distance

As we move our bodies in the world around us, we constantly enter into negotiations with others in physical space. These negotiations are typically based on the zones of interpersonal distance defined by Edward Hall[2]: intimate distance, personal distance, social distance, and public distance. In our daily lives, we see these distances as territories. Consider your own personal space, for example. Personal space has boundaries which we readily define and adjust. When our bodies move, our personal space moves with it. And, when threatened, we will protect our personal space in covert and overt ways.

Negotiation of interpersonal distance often begins with an act of invasion, the purposeful or accidental intrusion of a person moving from a more distant zone into a closer zone. Invasions are noticeable when another person moves from public distance into social distance, from social distance into personal distance, and from personal distance into intimate distance. However, our reactions to invasions of these zones are dependent on a number of other factors that influence our perceptions of the invasion, many of which emerged in the subway car example. These factors include spatial arrangement in the environmental context, the number of people present in the environment, the relative position and orientation of bodies, the physical characteristics of others in the space, and the identities of the various people involved.

Since Edward Hall's articulation of proxemics, numerous researchers have investigated the factors that mitigate our choices when negotiating interpersonal distances. Hartnett, Bailey, and Hartley identified elements of body position, height, and sex that influenced comfort in different zones of distance.[3] The role of height in interpersonal distance was reexamined by Caplan and Goldman using observations of passengers on commuter trains.[4] Remland, Jones, and Brinkman offered observations about interpersonal distance and body orientation as they related to culture, gender, and age.[5] Still others have investigated the physical arrangement of our bodies in motion. These give particular insight into negotiations of interpersonal distance as they occur.

For example, when we walk together, we space ourselves in patterns and arrangements that illustrate information about who we are in relation to others in the group. In the *Journal of Nonverbal Behavior*, researcher Marco Costa reported on findings related to interpersonal distances in group walking.[6] His observations of over 1,000 groups walking on an urban sidewalk suggest that the ways we position our bodies while walking in groups is related to size of the group, the composition of the group, and the characteristics of members of the group. Dyads and groups consisting of three, four, and five people were categorized by gender composition (e.g., all-male, mixed, all-female) as well as by similarities and dissimilarities particularly related to height. Based on these categories, Costa assessed the alignment of bodies in a group while walking and the group's walking speed. Interestingly, all-male groups walked fastest and exhibited the highest interpersonal distance across all group sizes. Mixed-gender groups walked slowest and exhibited the least interpersonal distance across all group sizes. These findings underscore the cultural and personal influences that exist in the study of proxemics. When we walk down a street with a group of people, we rarely consider how to best distance ourselves from other group members or how fast we are walking together. However, these markers of culture exhibit themselves in observable and often predictable ways. On the issue of group alignment, Costa notes that the arrangement of group members in space could reveal elements of group cohesion and interpersonal bonding.

Interestingly, other researchers have assessed the ways we extend our zones of distance to include others. For example, personal and social distances tend to extend to and alongside others in our groups. In a dyad, two people walking down a street are unlikely to allow a stranger to pass in between them, moving together to preserve their shared social distance boundary.[7] Dyads will attempt to stay together when an intruder approaches by tightening closer together, reconfiguring themselves into a single file pattern, or verbally addressing the perceived intruder. This expansion of the boundaries of social distance to surround a dyad was most pronounced in male-female dyads walking together and least pronounced in male-male dyads walking together.

Conversely, a lone pedestrian is unlikely to try to walk between a dyad on a sidewalk. The dyad seems to have an enhanced boundary of personal distance. The same principle of boundary expansion applies to groups at social distance. Researchers at the University of Wisconsin–Green Bay studied group deflection: the concept that the boundaries of group social distances could actually repel other individuals.[8] Interestingly, they observed that as a group increased in number, deflection increased. For example, a single passer-by would distance himself farther from a larger group than a smaller group. The group's social distance boundary increased in size as the group increased in size, and this boundary acts to deflect potential invaders approaching the group. At UCLA, another researcher explored this group boundary and its relationships to familiarity of group members and the density of the crowd in shopping malls and suburban sidewalks.[9] A later study in Turkey continued this discussion through the observation of ATMs, exploring the effects of crowding on personal distance.[10]

All of these studies and other related research demonstrate the complexity in and volume of factors present in the study of interpersonal distance. Digital technologies complicate the study of proxemics interactions not only by increasing complexity but also by intersecting with multiple factors. However, the empirical study of the impacts of digital technologies on interpersonal distances in physical space is strikingly limited. One example of the type of study that would shed light on this complexity was conducted by Patterson, Lammers, and Tubbs.[11] In this study, researchers studied the effects of mobile phone usage on pedestrian encounters. They employed a method similar to the group walking concept above, observing the interactions of pedestrians as they passed one another on sidewalks. Instead of distancing patterns, they observed looks, smiles, nods, and greetings. Pedestrians using mobile devices were less likely to smile at other pedestrians while walking. Notably, females using mobile phones were less likely to offer greetings than both male mobile phone users and people (male or female) not using mobile phones. This type of research and further replication of previous studies on interpersonal distances can describe the effects of digital technologies on interpersonal distances in physical spaces. We have a lot to learn here about proxemics and our changing culture.

In the meantime, using Hall's categorization of sensations present in interpersonal distances,[12] we can also lay a foundation for further exploring the approximation of interpersonal distances in online spaces. When Edward Hall discussed interpersonal distances, part of his categorization of distances was grounded in the way that people communicated inside each distance zone. In particular, he emphasized volume, facial expressions, and linguistic styles as well as the closeness of bodies, measured through sight lines and the amount of discernible visual detail allowed at a given distance. Digital communication has the ability to alter the ways these distances are perceived in space and to recreate the zones of interpersonal distances for shared spaces created online. However, note that Hall's categories of sensation (kinesthesia, thermal reception, olfaction, vision, and oral/aural) are not all replicated by encounters in digital technology. Visual and oral/aural perceptions remain the staples of online interaction.

Anecdotal observations of mobile phone use might persuade us that people attempt to approximate social, personal, and intimate distances when communicating digitally. Consider video-based conversations between two people on mobile devices. When participating in a video conference for work, the camera is usually placed so that the speaker's head and shoulders can be seen at once, giving an approximation of social distance. This might also mean creating a more formal camera position by setting the camera or device upright on a desk or workspace and controlling the background information. When speaking to a friend, we allow the camera to move closer to our bodies, often showing the whole face with the head's perimeter discernable at the edges of the screen, a digital approximation of personal distance. Interestingly, in this scenario we also allow the camera to move around our environment with us, jostling the image, but creating a more informal tone reminiscent of the linguistic patterns at personal distance. With a romantic partner, we might bring the camera closer so as to fill the frame with part of the face showing our eyes, nose, and mouth, but cropping out the perimeter of the head, ears, hair, and neck. At this distance, the subtle details of the face are discernible, in a digital approximation of intimate distance. While we can change the visual composition and camera angles on screen, the visual dimension currently remains our primary mechanism for negotiating distance. The absence of the remaining sensations leave a chasm between digital approximations of social distance and digital approximations of intimate or personal distances.

In addition, the sharing of both intimate and personal distance also involves haptics, the ability to touch. At personal distance, one person can easily reach out and touch another, and incidental or accidental touching often occurs. At intimate distance, physical touch likely occurs more frequently than not. Two bodies close enough to sense body heat will likely touch each other. Obviously, in digital space, the sharing of intimate distance data is not connected to the opportunity for touch. Novelist Will Self describes it this way: "We spend our days surrounded by busily tapping fingers and hotly beating hearts, yet remain coolly inviolate,"[13] separate from the touch of others.

The absence of the ability to touch removes us from Hall's notion of intimate or personal distance quite abruptly. Moreover, the boundary between digital versions of personal and intimate distances remains unclear. Where and how designers find an approximation of personal distance is yet to be seen. Personal distance in the physical world requires elbowroom—enough distance to be close, but not too close. Providing elbowroom is a conundrum in this scenario, and I wonder if the distinction between personal and intimate spaces will even be necessary.

So, by Hall's rubric, when a dyad comes together through current digital technology, they are perhaps most likely to do so in a digital version of social distance. This claim is partly a function of the distance we perceive when interacting through technology remotely. At the same time, it is a reflection of the limits of the ability of technology to allow us to perceive finer details or certain sensations. For digital media to move into negotiations in personal distance and intimate distance, sensations beyond vision and oral/aural must enter the conversation.

This issue, raised earlier as a brief notation, illustrates that Hall's list of sensations that help define proxemic negotiations are not all replicated by current means of interaction via digital technology. Notably, kinesthesia (which Hall grounded in the ability to touch), thermal reception, and olfaction are not typically available in the context of digital interactions. But what if they were? Willy Wonka's smell-o-vision[14] comes immediately to mind. This imaginative tool most recently paired with the Roald Dahl fantasy sounds absurd—passing smells electronically from one location to another—but it might offer a glimpse into the future of sensing technologies.

Current researchers seem more interested in tracking thermal reception (body heat) than olfaction (smell), but the principle is the same. In one study in the DUB (design:use:build) group in human-centered design and engineering at the University of Washington,[15] designers experimented with temperature sensing and sharing technology. Participants used prototypes of thermal sensors and shared their experiences with researchers to generate a list of possible benefits and concerns related to temperature sensing. The participants suggested that thermal sensing could benefit remote relationships by promoting intimacy and coordinating conversation. For example, participants in the temperature-sensing trials noted that whereas thermal sensing in the bedroom would be acceptable for remote romantic relationships, thermal data sharing might be less appropriate for friends and family members. The ability to see the routines of a remote romantic partner through thermal sensing is at once reassuring and perplexing. DUB researchers Chang, Hwang, and Munson concluded that contextual information might have to be provided alongside thermal data for its best use. Thermal sensing, by Hall's definition, is the province of intimate space.

In another study, Rain Ashford at Goldsmiths, University of London, designs wearable technologies as broadcasters of emotive displays—what she calls emotive technologies.[16] Children of the 1970s might think of these akin to a high-tech mood ring. While measuring physiological changes in the body, Ashford's Visualising Pendant broadcasts colors or shapes to amplify other nonverbal signals of attentiveness, boredom, or relaxation. A person in conversation with the wearer of the Visualising Pendant might see the pendant suddenly display a bright red letter X causing him to change the topic of conversation, and then note that the pendant changes back to a subtle green. Ashford says the goal of the pendant (and likewise the goal of her Baroesque Barometric Skirt) would be to disrupt the typical flow of nonverbal communication. Like the thermal sensing technologies above, users would lose a bit of the backstage privacy allowed in typical social situations in exchange for amplified nonverbal cues.

My reading of these findings in relation to proxemics is that the perceptions we expect to experience in a negotiation of personal and intimate distances—kinesthesia, thermal reception, and olfaction—might be currently perceived as too intrusive for digital sharing. One of the issues previously noted is that information shared digitally is also shared publicly. Thus, if we do not want information from intimate distance shared publicly, then we may also not want it shared digitally. On one hand, negotiations of intimate distance shared online with romantic partners might create feelings of closeness and intimacy, but negotiations of intimate distance shared online with the general public might seem an invasion of privacy, or just TMI.[17]

These descriptions indicate that technologies will be able to recreate the remote sharing of thermal reception, if not also later olfaction and kinesthesia, as a digitized form of intimate distance perception. Currently, these types of technologies remain in trials without clear application or adoption. Thus, digital media arguably collapse most of the negotiations of interpersonal distance into a single category. By virtue of the visual and aural components of the technology, intimate, personal, social, and public typically become social distance when digitized. Perhaps it is for this reason that participatory media is labeled, by definition, as social. As a refresher, these are some of the characteristics of social distance negotiations[18]:

- Head size is perceived in normal size (not obscured as in intimate distance, enlarged as in personal distance, or diminished as in public distance).
- Details of skin, eyes, and other facial features can be focused upon with a moving gaze, but fine details are obscured.
- Conversations use normal vocal volume.
- No expectation for touch exists, and touch would require some deliberate effort.
- Social distance allows people to be screened from one another.

Because almost all of our negotiations of interpersonal distance in shared online spaces occur at social distance (at least for now), users of those spaces might do well to use the same behaviors and information-sharing techniques used in shared online spaces that they would use in physical conversations at social distance. For this reason, social space warrants further attention as a design component of proxemic interactions and interpersonal distance negotiations.

Sociospatial Design and Digital Technologies

Digital technologies that alter our interactions in and with spaces have the capacity to alter and control our behaviors. In terms of sociospatial design, digital technologies can be characterized as sociofugal or sociopetal in their abilities to help us distance ourselves in relation to one another. Even the concept of classifying digital interaction on this dialectic has the potential to upend our discussion of proxemics. Nevertheless, it is a valuable thought experiment to examine.

Sociofugal Digital Technologies

Like chairs placed back to back in a living room, digital technologies can also screen us from one another to create sociofugal designs. When comedian Patton Oswalt described what he learned by quitting Twitter for a summer, he discusses the direction of his gaze. Because he wasn't looking down at his phone, he instead looked around at other people who were looking down at their phones. "Sometimes I'd catch the gaze of a holdout like me. A freak without a phone. Adrift in this gallery of bowed heads."[19] In his depiction of a technologically mediated lifestyle, people using their mobile devices peered down into technology whereas people who, either by choice or necessity, stowed their phones saw the world and the other people around them.

This issue of the sociofugal nature of diverted gaze has become a common concept throughout popular culture. Our attachment to mobile phones has been a growing part of plot lines in romantic comedies, action flicks, made-for-television dramas, sitcoms, and everything in between. The parent complains about teenagers texting at mealtime. The family travels to a secluded lake with no cellular service. The child longs for her workaholic mother's attention. The cracked mobile phone left lying on the ground serves as a remnant of a friend in dire need of help. Our attention to our devices takes our attention away from other things. Let's pause to note that this is not necessarily a new idea. Similar issues were raised with books, radios, and televisions. Of these, the book and the mobile device share the most in common as sociofugal devices. At the onset of radio and television, the devices were used as collective entertainment. Families gathered together to listen or watch together. Books and mobile devices are generally individual technologies enjoyed by a single user.

In the case of the mobile phone, the act of receiving a call can alter one's relationship with others in space. Among the Bulsa, a people of upper east Ghana, the addition of the mobile phone to their African cultural lifestyle indicated some nuanced and intriguing findings about humans in relation to digital tools.[20] Whereas a physical interruption of two people in face-to-face conversation would be largely forbidden in Bulsa cultural contexts, the virtual interruption of the cell phone is widely permitted. "Cell phones invite Bulsa to make a choice between attending to virtual or face-to-face conversations and relationships. Rather than offering a free choice of the alternatives, however, cell phones stack the deck, so to speak, in favor of the virtual."[21] This indication of the sociofugal nature of mobile devices in face-to-face interpersonal contexts could be read as the alienation of individuals or the democratization of individual access and is open to other possible readings as well, raising compelling questions about the benefits and consequences of technologies as sociofugal devices.

The loss of connection between people and nearby others remains one of the most repeated lamentations voiced about digital technology. My family and I were walking through a picturesque park in our town and wanted to take a picture of our whole group. At the time, we had two viable options: Ask a passerby to take a picture for us or use the self-facing camera on the iPhone to shoot a selfie. We chose the passerby. But, as we continued through the park we looked around to see several couples, groups, and families taking photos, and all of them were shooting selfies. As a child, when my family traveled to a tourist destination, we would frequently ask random strangers to take photographs of our family, and to reciprocate the favor when the same request was made. As a traveler, I remember relying on the camera to provide opportunities for interaction with both place and culture. Asking a local Frenchman to take a photo for me at Sacre Couer in Paris gave me an opportunity to inquire about nearby attractions, ask for directions, hear about the culture, and put my years of studying the French language to the test. The camera increased my opportunities for chance interactions with people I would never otherwise meet. Opportunities for those types of interactions are dwindling. In our current tourist destinations, the selfie has become the modus operandi of the day.

On the BBC radio program *A Point of View*, author and broadcaster Howard Jacobson called this trend "the tyranny of the selfie."[22] The selfie, and the selfie stick by extension (pun intended), propagate our self-reliance and our appreciation of our own viewpoints over those of others. "There's a subtle difference," Jacobson argues, "between having someone take your photograph and taking it yourself. A third party will see something you don't."[23] The selfie is an advertisement—a portrayal of self through the eyes of the self—whereas the third-party photograph is more journalistic in nature—a portrayal of a moment in time captured through the eyes of another. In terms of sociofugal design, the selfie stick is one small bit of technology that pulls us away from interactions with others, calling us to focus on ourselves. And the selfie remains a small but powerful example of the sociofugal power of digital technologies.

Remember that sociospatial design incorporates both real and imagined boundaries: the tangible fence as well as the intangible property line. In my observations, mobile phone use crystallizes a perceived boundary of personal space and deflects others from interaction. In daily life, when we board a subway and purposefully put in headphones or gaze at a mobile device (or a newspaper for that matter), we use the device as a nonverbal signal that we do not want to be bothered. However, we may also be attuned to recognize that same signal—I don't want to be bothered—when we are not purposefully trying to tune others out at a family dinner, doctor's appointment, work meeting, or child's birthday party. I wonder if our attachment to mobile phones simply prompts us to look whenever we are mildly bored. Perhaps onlookers read, "Don't bother me," when all we really want is to connect with someone. That would be a vicious circle created by a learned (and perhaps misguided) sociofugal interpretation of digital technology and personal space.

Sociopetal Digital Technologies

Facebook's mission statement reads: "Founded in 2004, Facebook's mission is to give people the power to share and make the world more open and connected. People use Facebook to stay connected with friends and family, to discover what's going on in the world, and to share and express what matters to them."[24] Share. Connect. Discover. These are the verbs of interaction: people facing people, 1.4 billion people facing 1.4 billion people and counting.[25] Like much of social networking, Facebook brings people together in real time in digital space, allowing active daily users to feel connected to each other. One might argue that social networks were founded on the principle that digital technology can and does create sociopetal online environments. As a scholar, I've been particularly interested in ways in which digital technology might create or facilitate the opportunity for community. Although I am only now labeling this as sociopetal, several projects have investigated this similar question: Can technology bring people together in sociopetal arrangements in the same way that a conversation in a coffeehouse or living room can?

My first academic foray into this conversation was with a bunch of geeks.[26] Over the course of several months, I captured conversations in public chat rooms. For the record, chat rooms were the popular online interaction site du jour in the mid-2000s, just as Facebook was beginning to transition from a college-only platform to a more public one. This study of these conversations between regulars explored whether people coming together in chat rooms might exhibit characteristics similar to people congregating in subcultural gathering sites like concert clubs. Indeed, the interactions between self-described geeks in the online chat rooms demonstrated a variety of elements common to other subcultures, among them resistance, conspicuous consumption, and style. At the time, I was particularly interested in the use of bricolage for visual representations of self on screen. Bricolage refers to the adoption of elements of mainstream culture as representations of affiliation with a subcultural identity.[27] Among the geeks in these chat rooms, avatar and screen name selection were the primary stylistic components on display. Each person in the chat room selected an avatar image to accompany their chosen name. These avatars were almost all borrowed images from popular culture representing anime, science fiction, or comics. In relation to proxemics, the point is that the chat room brought together like-minded individuals for discussion and collaboration. Their coming together occurred in real-time but spanned distance. A virtual social space for group interaction was created.

When I set out to study another digital and physical merger, Twitter use in the classroom,[28] I was particularly interested in student perceptions of instructor behaviors. Often these are identified as teacher credibility (the perceived level of expertise in the field studied), teacher content relevance (the perceived level of connection between course material and real-world application), and teacher immediacy (the perceived feelings of closeness between instructor and student). I expected Twitter use to create a connection between the class material and real-world events, judging that students who interacted with their instructors online would perceive higher levels of content relevance. And, indeed, when students reported actually reading the instructor's tweets, content relevance increased as predicted. What surprised me about the findings was that teacher immediacy was correlated not with actually reading the instructor's tweets, but rather with simply thinking that the instructor was active on Twitter. The more active on Twitter the instructor was perceived to be, the higher the degree of perceived closeness between student and instructor. So, the students who thought the instructor posted more frequently on Twitter also felt closest to that instructor. The opportunity to connect made students feel more connected. For the purposes of this book, I'm not arguing (yet) that Twitter makes people feel closer to each other. Instead, this study points out the finding that people can achieve higher immediacy—increased perceptions of closeness—by engaging in social media together. (Just as an aside, this study was rejected by a leading journal in the field before being accepted in another. The rejecting reviewer wrote that he or she would hate for the journal to publish any paper that suggested Twitter could be beneficial for the classroom. I now argue that the reviewer in question missed the nuance of the claim.) Kristen Bostedo-Conway and I expanded this finding into another study examining the relationships between Twitter and nonverbal perceptions across different relationship types, and other researchers around the nation have deepened and sharpened these findings in their own research on the intersection of digital technology and classroom space.[29]

Later, Ashleigh White and I expanded this research on the sociopetal nature of social media studying the role of Twitter chats as digital third places.[30] The concept was that Ray Oldenburg's characteristics of third places[31] might be able to be observed in the text of Twitter chats. By studying over 3,000 tweets from three established Twitter chats, we argued that digital gathering spaces could become third places, but with a few caveats. The space of these places remains somewhat contentious. Even though people were conversing with each other in a sociopetal fashion, their individual locations and perceptions of time were their own. This dis-location of the chat caused us to ask where the third place was actually located. Was it on Twitter or on a desktop or mobile device? Was it on the Internet or in the content? Was the third place characterized by the coming together in time more than the space itself? We will return to some of these questions.

Many other researchers and thinkers have contributed to this ongoing conversation. James Paul Gee labeled interactive chat-based forums (as well as any physical or digital space in which like-minded individuals gather) as *affinity spaces*.[32] Kristy Roschke studied a sub-Reddit populated by gluten-free information to assess the learning that can emerge in digital information sharing.[33] Dawn Gilpin discussed online identity construction and sharing through microblogs.[34] A team of researchers at the University of Minnesota conducted a field experiment to better understand how member attachment occurs in online groups and teams.[35] Likewise, social media researchers Shannan Butler and Corinne Weisgerber at St. Edwards University have argued for the use of social media to create personal learning networks for advanced study and professional development.[36] In a personal learning network, a professional connects with experts in a field of interest through blogs, microblogs, and social bookmarking tools. By using these tools, one can develop a flow of incoming information as well as contribute to the ongoing conversation surrounding their field. These networks are constructed with a focus on information sharing, but quickly begin to promote personal connections—networking in the traditional professional development sense.

All of these studies point to the sociopetal nature of digital technology, although none have labeled it as such. When we sit around a dinner table, we arrange ourselves facing one another to engage in conversation. When we arrange ourselves toward our screens and (virtually) face one another, we are potentially doing the same thing.

A Problem of Degrees

I fully realize that in the above discussion of sociofugal and sociopetal digital technologies, some readers might think that I am suggesting that attention to hardware—the devices we use—results in sociofugal experiences whereas attention to software—the applications on those devices—provides sociopetal opportunities. But this assumption limits the conversation.

From an information design perspective, this concept was well articulated by interaction design theorists Jay Bolter and Diane Gromala in their book *Windows and Mirrors*.[37] Simply phrased, the computer interface is a window through which we see information, but it is also a mirror that reflects our culture back to us. In interface design, if the technology is acting as a window, then we see only the content. However, if the technology is acting as a mirror, then we see the delivery mechanism, the device, or even the user (us). Bolter and Gromala remind us, "We are moving toward a philosophy of design that acknowledges both the place of computers in the world and the importance of the body and physical environment within and around the interface itself."[38] This emphasis on materiality—the physical body and environment—is a burgeoning field moving through art, but yet to be widely accepted in the social sciences. The window and the mirror of digital technologies both act in the context of the body (user) and environment (space). When applied to social networking, for example, Facebook as the window is sociopetal—orienting us toward our network of friends—whereas Facebook the mirror is sociofugal—orienting us away from the people sitting next to us. When we work through the platform of Facebook to see each other, we orient ourselves together. When we use Facebook as a distraction to remove ourselves from those in physical proximity to us, Facebook becomes sociofugal.

In our discussion of digital proxemics, the gap between sociopetal and sociofugal is at once material and digital. The gap is a matter of degrees on a spectrum. Remember that sociopetal is not "good" and sociofugal is not "bad." Both are used in balance to achieve optimum experience. For example, in a recent study, June Furr interviewed people with physical disabilities to learn about their use of Facebook. People with visible disabilities reported that Facebook became a more sociopetal environment for them than they had perceived in the physical world. They discussed their choices to reveal information about their disabilities, and often referred to the site as an unusual environment that allowed them to be seen as a person first.[39] Even though the site is sociofugal in that it creates orientation gaps between the bodies of people in the physical environment, the benefits of these sociofugal patterns made the use of Facebook more sociopetal for these users than the analog environments experienced every day.

I wonder what Edward Hall would say about orienting oneself in a physically sociofugal way toward a virtually sociopetal experience. Much of our experience to date with digital technology has this effect, particularly in social networking, but even more so in the rapidly expanding worlds of geo-locative applications, augmented reality, and virtual reality, which will be addressed more fully in coming chapters.

Our expectations of proxemics, interpersonal distance, and cultural tendencies in physical and digital environments can be employed to manufacture particular experiences for us. In fact, well-designed spaces often contradict or violate our expectations of interpersonal distances. Ubicomp designers Nicolai Marquardt and Saul Greenberg note: "Violating a person's personal space as defined by proxemics might not always cause a negative reaction. Depending on a ubicomp application's context and design, deliberate violations of personal space (such as requiring people to stand close to each other) might be an integral part of the user experience (for example, in games or public interactive-art installations)."[40] This wonderful notation in their research belies the fact that not only do we allow spaces to enhance and create experiences for us, but we also allow spaces to control our experience. Moreover, spaces and their incorporated technologies are designed to do just that.

· 4 ·

BODIES IN MOTION

Waiting in line is no fun. Unless you're in Walt Disney World. Just when the line-up is getting boring, a costumed staffer comes to take pictures with all the kids, or a playground emerges around a hidden corner, or a visual display lights up as an animated character launches into a monologue of jokes. The art of the queue is one that Disney World has advanced overtime, inspiring people to stand in line for popular attractions and also to (sometimes) enjoy their time spent waiting. By some accounts,[1] Disney has been so adept at queue formation that Disney's Imagineers have become the standard bearers of experience design in contemporary queue theory, the study of the mathematical construction of waiting lines. I once became so enamored with a queue area at Walt Disney World that I blocked the queue for several minutes to participate in the experience. Why? The intersection between bodily movement and digital technology caught my attention in an unexpected place and in a way I had not experienced in a queue before.

In late fall 2012, Walt Disney World initiated a soft opening for its new Fantasyland attractions including "Enchanted Tales with Belle," featuring the live-action heroine and characters from *Beauty and the Beast*, and "Under the Sea—Journey of *The Little Mermaid*," in which Ariel, the title character, recounts her story through a ride featuring the ocean floor, populated with animatronic characters.[2] On the day my family visited the Magic Kingdom, these two attractions were unexpectedly open for viewing and testing. Both were fun, engaging, and provided very different but equally unexpected experiences. My experiences in the queues for each were polar opposites. The Belle queue that day was long and wait times were high as the experience was tested, paused, and reworked over time. Because of this, the queue offered very little of interest to report and we waited a long time. Meanwhile, the Ariel ride shut down altogether during testing, causing the long queue that had formed to dissipate. We got in line minutes after it reopened, and there was no wait.

Even though there was no line to contend with to board the Ariel ride, park-goers had to navigate the circuitous, serpentine queue area to get to the ride. The queue design moves riders through a beautiful underwater-grotto-themed labyrinth that disguises line length and engages riders' spatial awareness. Embedded in the decorative stones in the grotto queue area is a series of small, glass, aquarium-like indentions. When I paused to peer into one of the aquariums, a digital crab peeked back at me and then ran into hiding. He quickly emerged with an object, willing me to play a matching game with him. It soon became apparent that not only did my presence trigger his, but also my hand gestures controlled his actions. These digital crabs compelled me to stop what was at the time a very fast moving line to play with the technology. Repeatedly. Much to the dismay of those in line behind me. The interactive crab as a motion capture device connected my body to my lived experience in this queue through digital sensing technology. My hand waved off objects or pointed to them, helping the crab decide how to organize his treasures. That crab controlled my movements. The seemingly simple but deceptively complex interaction represented a new digital intersection between queue theory and experience design. And it offered a digital-physical connection between my body and the technology on site.

Interactive digital crabs like those in Scuttle's Scavenger Hunt[3] that respond to gesture are minimally invasive, and represent only the beginning of a consumer-oriented trend to harness the work of our bodies and alter our bodily motions using digital technology. In September 2014, Apple CEO Tim Cook announced the company's much anticipated movement into wearable technology. Apple Watch, released for consumers in April 2015, is an attempt to resurrect the smart watch, a genre of wearable technology that began with a calculator watch in the 1970s. Apple Watch combines the industries of personal fitness, personal health information, personal banking, and personal communication. Note the keyword "personal" before each industry. The Apple Watch and its competitors are an invasion of the personal space.

"It's intoxicating and also a bit disconcerting to have this much functionality perching on your wrist, like one of Cinderella's helpful bluebirds," write Lev Grossman and Matt Vella for *Time Magazine*'s cover story on wearable technology.[4] They continue: "We're used to technology being safely Other, but the Apple Watch wants to snuggle up and become part of your Self." The Apple Watch—in conjunction with your iPhone—purports to carry your credit card information, manage your calendar, help answer your messages, and sense your heartbeat. These functions shouldn't surprise even the newest smart phone user, as our devices have been engaged in this activity for the last decade, if not longer. However, we've never strapped them to our bodies and allowed them to move with us until very recently.

Reporting in a later edition of the same magazine, Bryan Walsh described the digital tracking he employs to monitor his body, including his nutrition, sleep patterns, and training plan for the New York City Marathon.[5] Using wearable technology, Walsh and many others are investing in the quantified self—the movement to report information about the body in numeric form in hopes of personal improvement. Self-monitoring is not a new phenomenon, but wearable technology is. The question is whether or not this digital technology (and the information it reports) changes the ways we move.

In 2010, Bonnie Rochman[6] reported on the use of simple messaging system (SMS), also called text messaging or texting, to advance prenatal care for expectant mothers. According to the *Time* article, text4baby provides enrolled mothers with texts scheduled in relation to their due date. Texts might remind mothers to see a doctor early in the pregnancy, to make behavioral changes surrounding nutrition and body care, and suggest practical just-in-time tips, like how to position a seatbelt around a pregnant abdomen in the third trimester: "Shoulder belt goes between your breasts & lap strap goes under your belly (not on or above). Wear it every time."[7] These health tips aim to suggest behavioral changes about the body. Surely in the past, women could have sought this information from peers, books, or Internet searches, but the ability of an organization to change a woman's interaction with her pregnant body through text messaging changes the fundamental nature of information seeking and sharing. Users no longer need to seek the information, they only need to be mindful of information when it arrives.

Whether interacting with a digital crab through motion capture, using a wristband to collect and store personal health data, or allowing text messages to teach body positioning, accepting or rejecting digital cues about how and why to move the body situates technology as a benevolent user aid for the body. And it may very well be. I've used applications on my iPhone to train for a 5K race, time labor contractions (my wife's, not mine), receive reminders about friends' birthdays, and chronicle nutrition information. Willing participants request the support of these aids by enrolling with services that provide text messages, alarms, wearable alerts, or other reminders. We ask digital technologies to inform us about how, when, and where to move and, in many cases, we rely on them to offer reminders, rewards, and even encouragement. Over time, we believe that we assess and manage the impact of these technologies on our bodies. But, I'm not so sure that we actually do.

I remember an interaction I had with a friend wearing Google Glass. His appearance was normal. Behind the glasses his eyes and facial features were the same as they had been the day before, without the glasses. Yet, he was different. His eyes moved differently than they usually do. At times, his eye contact was the normal, predictable gaze to which I was accustomed. But at other times, his eye motions became confusing, even erratic. His eyes blinked at odd intervals. He looked up at icons unseen to me, but when I followed his gaze upward, my glance revealed only clouds in the blue sky above.

I think about Google Glass from time to time. As I don my eyeglasses, I consider the ways they help me see the world—to make it more clear, to make my vision more acute. I wonder if Google Glass will serve the same function. By overlaying digital messaging onto the physical world, will our ability to see increase? Or will we see less? The ways our bodies move are fundamental to our lived experience. Beyond its physiological importance, motion of the body directs our communication with others, it defines our relationship with the space around us, and it gives others a glimpse into our culture. Herein lies a fundamental question for our time. What does technology add to our bodies and what does it cause us to lose?

Extending Man

Technologies from hammers to cameras and automobiles to writing instruments have changed the ways we carry and move our bodies in space and time. Fifty years ago, Marshall McLuhan penned *Understanding Media: Extensions of Man*[8] in which he re-imagines media[9] by addressing them as technological extensions of the human body. He examined 26 specific media in separate chapters to elucidate his arguments around the role of media in our lives. Many of McLuhan's thoughts on media (e.g., *the medium is the message*) have become and remain part of the growing lexicon of new media. McLuhan's observations about media in the 1960s have become increasingly clear in light of present-day digital media and as media and the messages they present become inextricably intertwined.

McLuhan states that, prior to Cubism, the message of a work of art was its content—an assertion indicated by viewers asking what a painting was about. Conversely, Cubism presents the whole of an object—front and back, inside and outside simultaneously—in two dimensional space. For example, Braque's *Woman with Guitar*, 1913, illustrates all sides of the subjects on the same visual plane. Duchamp's *Nude Descending a Staircase No. 2*, 1912, integrates time and motion as multiple perspectives in two dimensional space. The human subject is seen in multiple positions, simultaneously, as it moves down the staircase, resulting in what contemporary viewers might read as a blur of motion down the staircase or multiple images superimposed atop one another. McLuhan suggests that Cubism ushered in new thinking about the intertwining of medium and message. In his view, medium and message are simultaneously presented and read together. Electricity and digital technologies continue to further this idea and complicate it for both media designers and users.

McLuhan's importance to our study becomes clear in the notion that man is connected to the medium: "The message of the medium is the change of scale or pace or patterns that it introduces into human affairs."[10] The medium is not a separate entity to be assessed apart from the human context in which it exists. In the category of personal fitness, wearable technology can have implications for our health not only because we can keep track of our calories in (and calories out), but also because it alters the pattern of our behaviors while wearing them. Knowing that the FitBit is recording our movements changes our movements. It's the old Hawthorne effect for a digital age. Technology changes the ways we move our bodies.

Only two years following *Understanding Media*, McLuhan wrote, and lamented that "all media work us over completely."[11] His perspective suggests that, like park-goers in a queue, users are not controlling or reflecting on the media available to them. Media begin to blend together and man becomes connected to them—both in the designing stage and in the consumptive stage. Fifty year later, studies like the current one no longer ask if McLuhan's premonitions were correct. Now we only ask how this blending is occurring, and, perhaps more important, call for reflection on these connections.

Marshall McLuhan's writing and wildly visual depictions of man's interaction with technology[12] illustrate his views concerning the connection between technology and the movement of the body. The concept that media are extensions of the body begins to articulate the complexity of human experience in a digital world. Donna Haraway writes in A Cyborg Manifesto, "Communications technologies and biotechnologies are the crucial tools recrafting our bodies."[13] Perhaps no one will then be surprised that computers and digital technology alter our physiological movement. Research on the human body assumes that technology of all sorts changes kinesthetic practice. This assumption leads health researchers not to prove that change occurs, but rather to describe the physiological difference that can be attributed to computer use.

For instance, computer use alters the ways our musculoskeletal system operates. In a 2007 Public Health Reports article[14], Drs. Leon Straker, Peter O'Sullivan, Ann Smith, and Mark Perry surveyed and photographed eight hundred and eighty four teenagers to determine if a relationship exists between body posture and computer use. Their results indicated that the more time a teen spent on a computer, the more variation of their posture occurred. Surprisingly, the postural variations were completely different for males and females. The relationship between posture and weekly computer use in males revealed a tendency toward head and neck flexion whereas the same computer use for females resulted in lumbar extension. Not only do these postural variations occur differently in males and females, they impact different parts of the body. Head and neck flexion occurs when the head moves in a downward nodding motion. From a postural perspective, habitual head flexion is revealed when the head is moved in front of the shoulders. As I typed that last sentence, I was suddenly aware of my own posture, and straightened my neck. Conversely, lumbar extension is an arching of the lower back. Habitual lumbar extension is demonstrated by an exaggerated curvature of the spine. The researchers suggest that these gender differences in posture could be related to any number of factors including average height, social expectations, and computer activity patterns. These variables often present different challenges for posture and sight lines. A taller person would be more likely to lower the neck or head. For example, the researchers indicate that males' use of computers for non-educational games versus females' use of computers for social interaction (via email and social networking) could offer some patterns related to types of input systems used and resulting postural tendencies during computer use.

Some would read this research report and suggest that any number of technologies from the book to aquariums to writing with a pen could create similar postural variations. They would be correct. Digital technology is not unique in its ability to impact posture. The concept here is not that digital technology fundamentally changes our posture by its own nature. Rather, frequent use of computers and other interactive screens changes the way we habitually hold our bodies.

Beyond postural issues, other computer-related body disorders have been identified by researchers, mostly related to eyestrain, carpal tunnel syndrome in the wrists, and repetitive strain injuries in the thumbs and fingers. For example, a team of researchers studying the impact of tablet use on the wrist[15] found that tablet use promotes problematic wrist positions. They concluded, "Touch-screen tablet users are exposed to extreme wrist postures that are less neutral than other computing technologies and may be at greater risk of developing musculoskeletal symptoms. Tablets should be placed in cases or stands that adjust the tilt of the screen rather than supporting and tilting the tablet with only one hand." As I read their article on my own tablet, I started to question its impact on my body. In the study, the movement and position of shoulders, arms, and wrists were assessed alongside hand dominance and type of tablet. Type of tablet had no effect on body position, whereas hand dominance data indicated that the dominant hand and non-dominant hand were often in different positions. The dominant hand manipulated elements on screen while the non-dominant hand steadied the tablet. The dominant hand demonstrated higher mean extension, or more habitual bending of the wrist. This study and others like it suggest that, from a physiological perspective, computers can alter our bodies in generally negative ways, especially when we use computers and other hardware devices either incorrectly or for prolonged periods of time (or both).

In terms of the physiological movement of the body, this chapter is interested in the relationship between digital technology and the motion of the body. Beyond posture and extension and in addition to its role as a benevolent aid, digital technology's connection to bodily motion reveals itself through the digital-human interactions of motion-capture technologies.

Motion Capture

In the mid 1980s Paul McAvinney and his colleagues at Carnegie Mellon University were brainstorming ways to capture human expression in digital formats. Long before our world of touchscreens and gestural interfaces, their team and others elsewhere pioneered the field of optical gesture sensing. Optical gesture sensing allows a user's movement to flow smoothly from physical surroundings into a digital format. One of the team's responsibilities was to devise gestures that seemed intuitive for users. For example, his reports suggested that a two-finger pinch across a touch screen would be a useful behavioral gesture for zooming in on an image. This suggestion came at a time when most digital natives were in primary school, learning how to program an Apple 2e to draw a series of lines on a green monitor display.

McAvinney's goal was to capture the motion of the human body in digital form.[16] On a piece of glass, McAvinney worked to develop a visual system that could discern the position and velocity of a fingertip on glass. He pioneered this invention into a VideoHarp: a glass paneled box that reads the movement of fingers as if it were a harp. The player can pluck, strum, dampen, and create chords on the harp through finger placement and speed. McAvinney noted that by connecting different sounds to the equipment, a musician would be able to apply stringed sounds to non-stringed instruments: pluck a trumpet, strum a drum, or create chords on an oboe. Motion becomes sound.

VideoHarp technology was also used to create a room-sized VideoHarp that captured the bodily movements of dance and converted them digitally into sound. A dancer could learn how to dance a sonata or gyrate to produce dissonant chords. Although the VideoHarp was an attraction in several symphony performances, it never caught on for widespread use.

One of the major applications of this technology, according to McAvinney, is the extensor: a digital tool worn to capture the motions of the human body. An extensor in physiological terms is a muscle that, when contracted straightens or extends another part the body. McAvinney uses this term to represent a digital connector that could recognize the motion of the body and convert the motion into digital code.

The technology to digitally capture body movement exists and has been progressing in both military and entertainment applications. In a *Scientific American* article in 2009, Larry Greenemeier noted, "The armed forces have been using simulators to train troops and pilots for battle since World War I, but this virtual combat has typically relied on joysticks, visors and other sensors to make the action realistic. New developments in motion-sensing technology, however, promise to deliver training simulations that let a user or group of users control the on-screen action using their own arms and legs to communicate with the simulator."[17] OpenStage, the technology discussed in the article, uses camera triangulation to detect body position and place it in digital space.

In early animation, animators filmed people or animals in motion and then used the frames of the film to guide or even trace images for frame-by-frame animation, a process called rotoscoping. Disney animators famously used these techniques in *Snow White and the Seven Dwarfs*[18] and in the final dance number featured in *Sleeping Beauty*[19] to make the motions more fluid than their contemporary competitors in animation. This early innovation has been made both easier and more complex by digital technology. Many animated and live action films utilize motion capture to translate body movement into the images we see on screen. *The Lord of the Rings* trilogy (2001–2003)[20], *The Polar Express*[21] (2004), and *Avatar*[22] (2009) were noteworthy expansions of this technology, among many other examples, that raised the profile of motion capture in film upon their respective releases.

Use of this technology continues to raise questions for the entertainment industry about the definitions of animation. In 2007, the Disney/Pixar animated production *Ratatouille* included a notation in its credits sequence which reads: "Our quality assurance guarantee. 100% genuine animation! No motion capture or any other performance shortcuts were used in the production of this film."[23] Despite this inferred claim about the sanctity of animation absent of motion-capture technology, the practice and the technologies that capture motion are becoming more diverse and widespread. Markers attached to bodies can digitally capture motion through optical, inertial, magnetic, or mechanical means and display the movements in digital form.

Andy Serkis, an actor who is perhaps the leading motion-capture animation performer of the last decade, described the evolution of the technology over the last fifteen years for *Popular Mechanics*.[24] In Serkis's terms, the technology has shifted from the capture of bodily motion to the capture of an actor's performance. In 2001, Serkis played Gollum in the Lord of the Rings Trilogy[25] and then reprised the role in the Hobbit trilogy a decade later.

Serkis's description of the difference between the two experiences details the evolution of performance capture: "Because you can shoot on a live location, you can do it all in one hit, rather than splitting it up into the performance being rotoscoped or key-frame animated and then going back a year later. You're just in the moment and you're acting with the other people on set."[26]

In the earlier version of Gollum, the character was a blend of rotoscoping and motion capture during post-production. According to Serkis, each frame was shot twice, once with Serkis's performance on camera and once without. With two scenes, producers had the option of either painting the character overtop of Serkis's form on camera or inserting a motion-capture animation into the empty scene. This second process was sometimes completed over a year after the original scene was shot, with Serkis re-enacting these scenes in front of a green screen. When *The Hobbit* was shot, technology had progressed so that Serkis's role was almost completely live action, pinpointing his performance in the moment. This, he offers, is the difference between motion capture and performance capture: the ability to capture an actor playing a role rather than blending motion into a figure.

The technologies described above translate physical bodily motion into digital bodily motion. Taking McAvinney's concept of extensors and combining them with motion-capture technology, the possibilities begin to take shape for a digital interplay of sensory techniques. Could physical bodily motion be translated into digital sound? Digital taste? The production of tangible artifacts on a 3D printer? A piece of visual art? A multimodal production of sound and light?

Consider these two examples as visions of the near future of digital interplay in motion capture and sensation.

Bodily motion becomes sound. In cases of visual impairments, researchers at the University of Memphis have begun testing Google Glass as an aid for dyadic conversation. Engineers Anam, Alam, and Yeasin[27] designed an app using motion-capture technology to analyze facial expressions and relay auditory cues to the wearer. So, if I had a visual impairment, I could wear Google Glass and use this app while carrying on a conversation with you. Even though I might not be able to discern the variety of facial expressions you make, the app would send me messages about where you are looking, how your head is tilted, and if you were smiling or yawning, for example. Based on these nonverbal cues, I might be able to communicate more effectively by knowing if you were getting bored or if you were really engaged in the conversation.

Motion of the body becomes visual art. Daniel Keefe, director of the interactive visualization lab at the University of Minnesota, reported on the CavePainting Interface, which functions inside a virtual reality immersive environment.[28] The interface reads the artist's intent by updating the environment based on the artist's eye movements. As the artist moves, CavePainting captures gestural movements in open space to create a visual representation of the motion of the hand. Keefe notes that the full motion of the body can create complex visual representations in this interface.

These two examples remind us that the room-sized VideoHarp discussed previously was only a few decades ahead of its time. Digitizing the motion of the body offers practical and artistic benefits. And the technology to engage with motion capture is rapidly expanding. Whereas many motion-capture tools use handheld sensors to place the body in space, gesture recognition would eliminate the need to attach sensors to the body. In the journal *Artificial Intelligence Review*, Siddharth Rautaray and Anupam Agrawal conducted a meta-analysis of research into the improvement of hand gesture sensing in human computer interactions.[29] The review illustrated the wide interest in this particular subset of gesture sensing and identified issues of importance to business and personal-use practices. Real-time accuracy of gesture recognition continues to be the interface issue in this particular case. But these technologies are quickly developing. Microsoft recently reported on Handpose vision-based based hand gesture recognition for human computer interaction: a gesture recognition software with increased accuracy and faster processing times.[30] Others, like virtual reality frontrunner Oculus, are working on enhancing similar technologies. Furthermore, designers are considering ways for this technology to be utilized not only inside specific programs or for virtual or digital gaming,[31] but also for mundane and ordinary tasks like web browsing or answering the phone.[32] As motion capture continues to become more detailed and nuanced, the ability to translate it into other sensorial expressions will only continue to grow.

Why We Will All Become Robots

The expansion of motion capture into extensor technologies offers possibilities beyond the entertainment, gaming, and military applications. Paul McAvinney notes[33] that extensors could allow the human body to coordinate teleoperation (the manipulation of remote objects), bioengineering and medical applications (the manipulation of tiny or microscopic objects), transportation (the manipulation of objects in motion), and work (the manipulation of objects beyond human strength). He calls this the "why we will all become robots" argument—the trajectory that will lead us to modify our physical bodies with digital tools that not only capture and track our motions, but also respond to them.

When considering the meshing of man and machine, the immediate question we ask is how much is too much? Media storytellers and filmmakers have raised this question through the futurist science fiction of *RoboCop*[34] and *Iron Man*,[35] the dystopian futures concocted in *The Matrix*[36] and *Terminator*,[37] and the cyborg societies of *Blade Runner*.[38] These stories and the many others like them investigate the synthesis of man and machine, and the extreme, yet seemingly possible, outcomes of such action—with widely varying results. Some depict technology as a facilitator of super-human strength, others suggest technology as the primary instigator of the demise of humanity.

In *RoboCop* and *Iron Man*, technology allows humans to use it for their advantage, build physical strength, and prolong life. In *The Matrix* trilogy and *The Terminator* series, the human race exists at war with machines that seek to rule the world. Humanity's only hope lies with small groups of people that seek to destroy the technology they created. In *Blade Runner*, the cyborgs— machines created by humans to function as humans—are so human that they don't realize they are cyborgs. Everyone is confused, even the audience at times: Who is a machine? Who is human? And does it matter?

Even though we recognize that these films are extreme portrayals of possible futures, this tangential trajectory into bio-applications of digital technology is an unavoidable one. In academic conversations, techno-human interactions in the body and their implications for humanity have been given voice by many. Georges Canguilhem, a French philosopher, wrote about the "normal body" and the malleability of that term over time.[39] Donna Haraway, professor emerita in the University of California system, wrote about technologically modified humans as cyborgs alongside the pop-culture phenomena of the 1980s and early 90s.[40] N. Katherine Hayles, professor of literature at Duke University and literary critic, describes the cultural shift toward technologized bodies as "posthuman."[41] I share these examples to illustrate that key thinkers have been debating this topic for many years, but much of that conversation is only now becoming part of a more mainstream discussion of our present rather than our future.

Currently, we accept these fusions of man and machine when we make physiological modifications to our bodies for health benefit. The pacemaker consists of a computer implanted in the body to regulate heartbeat. Vagus nerve stimulators[42] perform a similar function in patients with severe epilepsy. Prosthetic devices incorporate digital technologies to aid in limb function.[43] Pet owners implant their pets with radio-frequency identification (RFID) chips for identification purposes and some people are beginning to adopt this technology for subdermal implantation in humans.[44]

Many, if not most, citizens would advocate for allowing a pacemaker-type of computer to be embedded in the body for life-sustaining treatment. But, those same advocates might flinch at the idea of replacing one's functioning arms with stronger, bio-digital limbs capable of heavier lifting, or amputating muscular legs in favor of titanium legs and joints that allow us to run farther and faster.

The case of Paralympian and Olympian Oscar Pistorius is an interesting study in the history of prosthetics.[45] Pistorius was born without a fibula in either leg (one of two bones of the lower leg below the knee). As a child, his lower legs were amputated and replaced with prosthetic legs. Pistorius mastered his new legs by age 18 months and became an avid sportsman, winning gold and bronze medals in the Athens Paralympic Games. Wearing Ossur Flex-Foot Cheetas, blade-like prosthetic running devices, Pistorius later competed against able-bodied athletes in the International Association of Athletics Federations (IAAF) Golden Gala in 2007.[46]

Pistorius' entry into this competition resulted in a review of the biomechanics of his prosthesis by the IAAF. The review determined that the ergonomics of Pistorius's prosthetic devices actually gave him an advantage over able-bodied athletes. A legal battle ensued in which the ruling was reversed. Pistorius later competed in the 2012 London Olympics as a representative of South Africa as the anchorman in the 4x100 meter men's relay. He also carried South Africa's flag in the Opening Ceremonies.

Pistorius is a success story for the benefits of prosthetics. But the investigation of the functionality of his prosthesis underscores our hesitation about body-altering technologies, digital or otherwise. Prosthetics like those Pistorius used in the Olympics rely on the body's musculature to control the function of the prosthesis. Adding electronics to prosthetic devices creates opportunities for many more interactions.

Every year, more digital technologies are being adapted for use in the body that would allow internal power sources and even the nerves and brain, alongside musculature, to control these devices. Exoskeletons under development are already being used to assist wheelchair-bound individuals in becoming ambulatory.[47] At a conference in Budapest, Hungary, in 2014, Amanda Boxtel demonstrated the results of her 3D-printed exoskeleton, designed and printed by Ekso Bionics and 3D Systems.[48] Despite being paralyzed from the waist down for 18 years, Boxtel's exoskeleton empowered her to walk. Boxtel travels and speaks about the benefits and beauty of bionics, and demonstrates her functional exoskeleton for audiences around the world.[49]

Researchers at the Centre for Neural Engineering at the University of Melbourne in Australia are experimenting with potential bionic solutions for hearing loss and epilepsy. In addition, they are already testing electronic vision technology that connects to the nerves through a retinal implant.[50] At the Center for Bionic Medicine at the Rehabilitation Institute of Chicago, researchers are developing prosthetic technologies related to motor manipulation and mobility. These include muscle reinnervation, rehabilitation techniques, and algorithm creation for prosthetic devices that remember and adapt to the wearer's muscle movements.[51]

The US Food and Drug Administration (FDA) recently approved the DEKA arm,[52] a bionic arm, for human use. The prosthesis is connected to the brain through a series of nerves and electrodes that allow the user to move the arm and hand to perform complex tasks involving multiple motor skills. The arm has been affectionately named "Luke," a reference to Luke Skywalker, hero of *Star Wars*, who was fitted with a bionic arm a long time ago in a galaxy far, far away. The DEKA arm is powered by a battery and controlled by an internal computer. It is certainly less functional than a human arm across all measures. Yet, it allows an amputee to recapture some of the function of the human arm and hand. Its dexterity greatly outpaces traditional prosthetic arms. The bionic arm has a promising future. However, the DEKA arm signifies the evolution of bionics toward widespread use in prosthesis. A bionic arm is an excellent solution for an amputee, but is it an excellent solution for someone with two functioning arms? If the functionality of the bionic arm one day surpassed the functionality of your functional human arm, would you consider an upgrade?

Reporting for *Popular Science* at South by Southwest (SXSW) 2015, Lindsey Kratochwill interviewed MIT Professor and bionics designer Hugh Herr.[53] Herr's team has advanced the field of biomechanics, developing bionics including a powered ankle-foot prosthetic that propels the walker forward.[54] Herr says that the creation of bionics is at once electrical, mechanical, and dynamic. The goal is to make comfortable prosthetics that receive instruction from the body and send information back to the body to function as well as (if not better than) natural body parts. Herr, a double amputee, envisions a world where even those with a mild disability might choose bionics as a solution and a future world where body sculpting involves a "menu of body manipulations" that extends far beyond tattoos, piercings, and plastic surgery.

In this world of bionics, the script on human-computer interaction begins to flip. Before bionics, we used our bodies to tell technology how to move. With bionics, we begin to approach a world in which digital technology can also tell our bodies how to move. When the body tells technology like the bionic arm how to move, such movement is facilitated by an electrical connection to the brain through a computer. What happens when that computer is hacked? Could a hacker tell a bionic arm to spill a cup of water, to use a gun, or to constrict an airway? Or worse, could the hacker use the computer to send signals back to the brain? If so, could a bionic arm controlled by the mind be used in reverse, as a bionic arm to control the mind?

I understand these examples could seem far-fetched and fairly disturbing. The examples seem like something from a futuristic sci-fi movie description: "In the near future, accountant Milo Spann is a test-subject for a new bionic arm. All is well until, unbeknownst to Milo, the arm starts to enact revenge on his ex-girlfriends." If Milo's arm starts controlling him, then the filmmaker gets to decide who the villain is. Is Milo the villain, having secretly plotted against his former lovers? If not, the culprit could be the arm's maker who programmed the arm to act in terrifying ways. Or, the culprit could be a computer hacker who logged into Milo's arm using his mobile phone's wireless relay. Or the culprit could be the arm itself, having developed a mind of its own.

One of my former colleagues would call this a tail-wagging-the-dog argument, meaning that we cannot make decisions and policies today based on the worst-case scenario for the future. And, I agree. However, we are approaching a time when digital technology not only *influences* our decisions about the ways we move our bodies, but also *controls* our choices about how we move our bodies in time and space.

Control. Digital technology controls the motion of the body. Simply writing those words challenges me to reflect on the trajectory of digital technology in the human experience. In many ways, bionics is only the latest (and most obvious) development of this sort. And as this chapter has demonstrated, the historical trajectory of digital-physical interactions between technology and the body is long and vast. Embedded technology has been subtly shaping our movements in space and time for quite some time. As we move forward toward rapid adoption of wearable technologies and bionics, I wonder who will pause to ask the questions about how these technologies dictate the motions of our bodies.

This reflective challenge can occur now, when we have the time and space to ponder it. However, a time is coming when people will allow technologies built into their bodies to dictate their movements in ways that we cannot yet anticipate.

If you're scoffing at the sentence above, look down at the FitBit on your wrist and ask yourself whether you've taken more steps today because it is there. Its simple presence is willing you to step, to exercise, to move. Now imagine a FitBit of the future that can not only visually remind you to move but also get your legs moving when you're short on your daily count. Surely, we say, "I would never allow a FitBit to control my legs and exercise for me."

But what if it could simulate exercise by stimulating the muscles in your legs? Would you allow it to do that? What if it could track your caloric intake without you having to enter your foods? What if it could measure your heart rate, shock it back into rhythm, and alert a physician if you were in need of medical attention?

My point is that the lines between life-sustaining digital technology, life-tracking digital technology, and life-controlling digital technology are rapidly diminishing in size and weight. When the digital and the physical interact—when we fuse man and machine—we give up some elements of control. The greater the degree of potential fusion, the greater the degree of potential control.

· 5 ·

FINDING OUR WAY

Several years ago, I attended a convention in New Orleans, Louisiana. Naturally, I arrived early and spent a pleasant afternoon preparing for my presentation at Café du Monde with a plate of beignets and a café au lait.

This convention was one of those fantastic academic juggernauts attended by 7,000 of my closest professional colleagues. The most salient detail of the conference needing to be addressed for the moment is the fact that all those people are hungry after long days of presentations. In a walkable urban environment like New Orleans's French quarter, that makes finding a restaurant with open reservations next to impossible within a short walk of the convention hotels.

Luckily for me, I'm part of a tech-savvy dinner group of early technology adopters. We gathered for dinner on the first night and began walking together into the French Quarter. One colleague logged into Yelp to look at diner ratings for nearby restaurants, finding a list of restaurants ordered from highest to lowest ratings. Another searched OpenTable for seating times and found a list of options in alphabetical order. A third posted a question on Facebook soliciting restaurant suggestions from her wide audience, crowdsourcing a quick unordered list of eateries. I opened Foursquare to identify what was around us, uncovering a list of restaurants based on geographic proximity. Between the four of us, 3 restaurants emerged on all lists. We dined at our top choice, two blocks away, Emeril's highly rated NOLA with a fifteen-minute wait for their largest round table in the center of the restaurant. The atmosphere was vibrant, the food delicious, and the company even better. I felt a twinge of guilt as other large parties from our conference inquired about wait times and were put on long wait lists or made reservations for later in the week.

As I recount that story, it strikes me as odd that we never looked at a street sign for navigation. We likely missed a variety of important and historic buildings worthy of mention. We darted through an alley, bypassing a busier sidewalk. We did not interact with the places in between the hotel and the restaurant. And we gave little attention to the stories present there. What can I say? We were hungry.

After dinner, we stowed our phones back in pockets and purses and took in the feel of the city, experiencing the sounds of a jazz band leading us to Jackson Square, the dark mystical alleys leading between buildings and under wrought iron balconies, and the bright, exposed light bulbs of Café du Monde beckoning to us to sit and stay awhile for, yes, more beignets.

In *The Hidden Dimension*, Edward Hall suggests that sensorial experiences are the fundamental, cultural patterns that define and shape cities in America. Furthermore, he calls for a rebirth of the distinctive city: "Most important, the rebuilding of our cities must be based upon research which leads to an understanding of man's needs and a knowledge of the many sensory worlds of the different groups of people who inhabit American cities."[1] In this lamentation, Hall speaks of the importance of the sensorial environment of the city.

New Orleans, Louisiana, like many distinctive cities, has an atmosphere all its own. The feel of walking down slate and cobblestone streets. The mixing sounds of jazz music and echoing laughter that is at once festive and sinister. The intertwining smells of dark roasted coffee, shrimp, the river, and party remnants. The sights of people marching down streets lined with stucco and wrought iron railings. The strange intertwining of French, Creole, Voodoo, Christian, and American cultural and religious expectations. The visual reminders of the aftermath of Hurricane Katrina and its role in the feel of the city. These sensations offer an experience of the city. And if you're a visitor to the city of New Orleans, these are the things you expect to experience, even though you may not feel comfortable there. The sensorial environment of a city causes us to distinguish one city from another in our memory. And yet, the sensorial environment of the city is fundamentally and inextricably changed by digital technology. Digital technologies change the way people move in, around, and through built spaces, cities, and urban environments. And built spaces, cities, and urban landscapes provide a valuable opportunity for exploring the role of digital technology in culture.

Proxemopetal and Proxemofugal Experiences

Edward Hall's discussion of proxemics in *The Hidden Dimension*[2] concluded, in part, with a treatise on the state of the American city. The city, he argues, is a direct extension of the nature of the proxemic interactions of man. In the design of cities, we can observe built artifacts depicting human use of space and infer the values assigned to different modes and forms of interaction. Cities are a projection of culture and a projection of self. Our movements through them indicate our values. At the time of Hall's book, American urban centers were being decimated by suburban sprawl. The decline of the urban condition, for Hall, signified a cultural shift in man's use of space. At the time of this book, urban centers are experiencing a shift of another kind, as urban areas are revitalized and renewed. Hall might argue, as I do, that this shift in value is reflected in proxemics.

The contrasts between two thinkers—Manuel Castells and Henri Lefebvre—and their approaches to the urban landscape has relevance for our discussion about cities and the ways we move in, around, and through them. Spanish sociologist Manuel Castells conceptualized urban spaces as a dichotomy of information and location. In what he calls the "space of flows," information can move through both time and space, allowing for interaction in real time.[3] This space of flows exists in opposition to the physical world, which Castells calls the "space of places," and in which information has no mechanism of flow across space or time. From Castells' vantage point, the information age has a way of changing our use of time and space in cities. Contrast this approach with that of French philosopher Henri Lefebvre. Lefebvre theorized that an urban space is not characterized by size, geography, built environment, or information flow, but rather by the combination of all of these acting together.[4] In his view, people and information flow into and out of urban spaces in time and space. This comparison between Castells and Lefebvre is, admittedly, largely truncated. But in digest form, the contrast between these two approaches illustrates the issues of presence in space and geographic location for our discussion of cities. Can an urban dweller be in a city but not of the city? Can an urban organization be of the city, but not in the city? Can a tourist visiting a city be considered a member of that city during the time spent there? The physical spaces of built urban environments and the information landscape of built urban environments combine to define and regulate our experiences in them. Furthermore, our behaviors in relation to both physical and digital spaces are markers of the dynamic changes in contemporary urban culture.

The quality of built urban spaces directly relates to the human experience in those cities. To best capitalize on the human experience, architect Jan Gehl suggests that cities ought to be lively, safe, sustainable, and healthy.[5] A lively city is one that invites walking, cycling, and sitting in the environment. A safe city projects a feeling of security within. A sustainable city utilizes pedestrian and mass transit for maximum mobility. A healthy city emphasizes public health as a community concern. Interestingly, all four of his concepts relate directly to the pedestrian. People like to walk around lively cities. When many people move around cities by foot at all hours, the perception of safety increases. Pedestrians using their feet to move through the public spaces and utilizing mass transit contribute to sustainability. Walking as a daily routine promotes personal health and, when multiplied across many individuals, impacts public health overall. As you can see, Gehl's answer to the problems of urban planning is to build cities for people, which also made a great title for his book. Gehl's *Cities for People* addresses the architectural and aesthetic components of urban spaces that enhance or limit the usability of that city for proxemics interactions by pedestrians. Naturally, his book is filled with images of architecture in cities all over the world that create opportunities for people to engage with urban environments, for better or worse.

Design principles of urban landscapes continue to inspire much debate. Historically, urban areas shifted in design with the advent of the automobile. Many quickly growing cities of the last century, like Los Angeles, Dubai, and Atlanta, privileged the experience of the automobile over the experience of the pedestrian. They look beautiful from the road but people walking through them feel quite small. In contrast, many European cities and American coastal, colonial towns built long before the automobile, have tried to integrate the automobile in more confined spaces originally built for pedestrians. Gehl says that this is an issue of scale: "The human body, senses and mobility are the key to good urban planning for people. All the answers are right here, encapsulated in our own bodies. The challenge is to build splendid cities at eye height with tall buildings rising above the beautiful lower stories."[6] Attention to the pedestrian experience is being realized in current re-makings of urban centers, exemplified in the large-scale rerouting of traffic around Times Square in New York City[7] to the smaller-scale success of narrowing Main Street in Greenville, SC.[8] The pedestrian is making a comeback—as are the cities that welcome them.

Noteworthy architect and urban planner Edmund Bacon noted this historical shift in his book, *Design of Cities*.[9] In it, he references the planned town of Vallingby near Stockholm, Sweden. The town was planned and designed based on an architectural model through which the designer oriented the buildings, roads, and parking in the town in an aesthetically pleasing way. Bacon notes that the town looks beautiful from an airplane, as it was designed to be seen. However, when realized as a pedestrian walking through the streetscape, Vallingby has little appeal to offer. Working on the design of the town as a cardboard model versus designing it from the vantage point of the pedestrian creates two different experiences. As you might note, Gehl and Bacon are both describing experience design for urban spaces. Architects and city designers have been contemplating experience design for millennia considering the constant interplay between form and function, asking both how will a building look and how will people use it. When we extrapolate their thinking to understand the impacts of digital technologies on the urban landscape, we must again turn to the experience of the user.

The choice of how to use our technologies to shape our experiences is always ours to make. In a recent study of urban dwellers in Tokyo, London, and Los Angeles, researchers Mizuko Ito, Daisuke Okabe, and Ken Anderson tracked mobile device users' behaviors using their phones, music players, credit cards, and other identifying information objects.[10] Their research identified three common patterns of inscribing places in urban spaces: cocooning, camping, and footprinting. I love these ideas.

To imagine cocooning, think of a person sitting on a subway train with ear buds in, totally engrossed in his mobile device. The authors point out that mobile media in various forms (including newspapers, magazines, or books) have a primary purpose to shelter people from social and spatial engagement in selected places. They noted that this sheltering occurred most often when people were either in transit or filling time between events. The cocoon offers an escape from physical surroundings and co-presents others with boundaries well-defined by constructs of interpersonal distance. Camping occurs when media and devices are used to set up a physical campsite in a public place. The authors identified cafes, libraries, and public parks as typical campsites, although any place with wireless Internet access that is convenient and

comfortable might be used. Whereas a cocoon shuts out the external environment, an encampment is selected because of the ambience and desirability of
the location, and often because of its proximity to food. Sometimes I think
that I might prefer my office to be a series of encampments all over town. On
a regular schedule, people could still find and meet with me, but my scenery
would change and flex with activity, season, and weather forecast. Whereas
camping and cocooning are both examples of personal media creating a place
for the user, footprinting is characterized by a user's interaction with localized
computers such as ATMs, reward and loyalty cards, and access cards for urban
services like toll roads, buses, or trains. These information objects working
together can track a user's movement in an urban setting, creating a path of
footprints through the city.

Cocooning, camping, and footprinting suggest that media users in urban
environments across cultures have certain typical practices that can be observed in urban settings. These practices are inextricably tied to mediated
experiences with personal and public media devices. So, when technology
changes, behavior follows. From a cultural studies framework, the role of digital technology in the deconstruction and reconstruction of physical location
has been a topic of much debate. Media theorist Joseph Meyworitz[11] suggested
that media have the ability to dislocate physical location from personal presence. He suggests that when we use media tools we dissociate ourselves from
our geographic location and enter a new other space. Media theorists Rowan
Wilken and Gerard Goggin[12] build on this concept, arguing that a place is
bound up in the location, as well as the material things, the meanings, and
the practices present in the location, and suggesting that the mobile device in
particular forces us to reconceptualize the connections between the spaces we
inhabit and geographic location.

As we approach the urban setting using experience design in the context
of proxemics, this chapter builds on those ideas. Through the lens of proxemics and sociospatial theory, I argue that both digital and physical settings can
be experienced separately, creating a proxemofugal experience, or they can be
experienced collectively in a proxemopetal experience—the culmination of
which may be realized in augmented reality.

When digital technology creates a proxemofugal experience, we get lost
in digital technology and forget where we are. Proxemofugal experiences
occur when we focus on our digital environment and isolate ourselves from
the physical one. The concept of cocooning described above exemplifies
this type of experience, but cocooning at the wrong times is often labeled

as distraction. In their study of five busy intersections in Manhattan, health education researchers quantified the nature of technology-related distraction among pedestrians.[13] The team observed over 21,000 pedestrians to study the effects of digital technology on pedestrian safety. They compared technology distraction of those crossing the street during the walk signal and those crossing during a don't walk signal. Whereas 1 in 3 pedestrians crossing on a walk signal were distracted, nearly half of those crossing during a don't-walk signal were distracted. The data demonstrate that distracted pedestrians were consistently more likely than non-distracted pedestrians to cross a street during a don't walk warning signal. By creating proxemofugal experiences that remove us from our surroundings, digital technology has the ability to decrease our alertness toward our environments. Sure, disconnecting from your physical environment is no big deal if you simply miss seeing butterflies and clouds floating overhead, but as soon as you step out in front of a moving car, your surroundings will reconnect with you in some very unpleasant ways.

Proxemofugal experiences have been realized by the general public as evidenced in campaigns to ban texting and driving, in homes where parents of gamers purposefully limit game time, in the romantic relationship where one person is recounting his day while the other is catching up on Facebook. But proxemofugal experiences are not all bad. We might have a need to depart mindfully from a space to gain perspective, to learn something, to pause an argument or halt a worry. Sometimes we need to get away from a space, to cocoon.

However, when we consider digital technology and built spaces from a design perspective—the "What can it do for us" argument—we tend to focus on proxemopetal experiences. Exemplified in augmented reality, proxemopetal experiences call our attention to a space through digital technology. They connect information and location and require a certain *dasein*—a being there. The proxemopetal experience creates a user who is not only physically present but also present temporally and mindfully.

I love the idea that digital technologies could forge a stronger relationship between user and place. That's one of the reasons I first set out to write on this topic. When we connect with places, we can deepen our knowledge of history, location, behavior, and self. When you walk into London's Westminster Abbey, Beijing's Tiananmen Square, or New York's Carnegie Hall, the sense of place you feel is not only from the grandeur of the sight, but also from the history of the site. You are walking where others have trod. You add your footsteps to theirs. The value you ascribe to that history is evidenced in the value you ascribe to the location as well.

Proxemopetal digital technologies allow users not only to be more informed about locations, but also to reinscribe meanings of place. In addition, these technologies force us to make decisions about how technology can either be employed to expand our control of our own experiences or to cede user control of those experiences from user to technology.

Augmented Reality and Reinscriptions of Space

Augmented reality, broadly defined, offers us a variety of opportunities to examine proxemopetal alterations created by digital technology and aimed at the user's experience in built environments and urban settings. In its multiple forms, augmented reality exemplifies some of the issues of proxemics raised in this book and the ways that digital technology captures, informs, alters, and controls our use of space. Augmented reality (AR) describes a view of the physical world with a layer of digital data superimposed upon it. From a user-experience design perspective, augmented reality might be utilized in five types of experiences that mediate a physical reality for a proxemopetal experience: visualization, search, simulation, play, and, of particular importance to our study of proxemics, navigation.[14]

Visualization as Augmented Reality

Augmented reality as visualization refers to the addition of a visual layer of data to the physical surroundings in real time. This is perhaps one of the most common understandings of the application of augmented reality tools to the physical environment, and the most readily understood. Almost every history of augmented reality points to the yellow first-down line on broadcast television football games as the earliest commercial application of augmented reality. When viewers see the virtual down line, now also accompanied by a virtual scrimmage line, they are seeing a visual layer of data added to the physical football field. One of the most complex examples of this in consumer practice to date is Google Glass. The eyeglasses are clear so that the user can see the physical world. Digital data are projected onto the glass surface so that information can be viewed simultaneously. When the data onscreen and the physical landscape fuse, augmented reality is created. So, scanning emails or sending texts would not count. Instead, augmented reality would be created if the glasses were being used for navigation in concert with a geo-locative application. These sorts of applications have to be built for devices like Google Glass and married with devices for use. Augmented reality devices headed to market include Microsoft's HoloLens and Sony's SmartEyeglass, and Apple is rumored to be developing device options that merge AR functionality with "fashionable" design for wearable devices in the future.[15] The wearable component of AR devices is one way to deliver an augmented reality experience to people, but current wearables are still only being worn by early adopters. In the interim, augmented reality applications continue to be developed to connect people and spaces using devices that people already carry. In one attempt at augmented reality, Google released Goggles Translate in 2010.[16] The premise was that users could scan texts in other languages with a smart phone camera. Goggles would then translate the text into the user's preferred language. The smart phone as pocket language guide allowed users to build a digital layer of translated text on top of a physical layer of original text. Google abandoned the application in May 2014, saying that Goggles, as a stand-alone app, had no clear use,[17] but added the feature to Google Translate as Word Lens in January 2015.[18]

Search as Augmented Reality

Augmented reality as search refers to the use of new organizational patterns for receiving results to a query. For example, in Barcelona, researchers at the Universitat Pompeu Fabra designed an augmented reality application for iPad that allows users to browse physical shelves of books digitally. By combining RFID technology with augmented reality application, the RFID system inventories shelves of books in real time. The augmented reality app allows users to view the shelf through their iPad and scroll through the book covers.[19] Imagine the applications of this information. A user unable to reach the shelf or handle the books could locate them easily in a library. In a bookstore, the available books on the shelf could be paired with sales data and user reviews via the app. Translate this same application to a grocery store and the shopper could identify where purchases are located in the store upon entry and if they are in stock. And so on. The researchers tested this same application, this time on a shelf of DVDs and using a smart glass interface, with similar results.[20] Their two studies reveal the connection between place, device, and application required for successful augmented reality development in the area of search.

As an example of applying the function of search to a physical space in Taiwan, an interdisciplinary group of researchers tested an augmented reality guide for tours at a heritage site. The guide used geo-located mobile devices to provide historical information about the site to users as they navigated the area. The application allowed users to visually zoom in on architectural details mentioned in the tour for closer inspection and presented data about the site organized by geography and proximity rather than other organizational patterns. When compared to an audio guide or no guide, the augmented reality guide produced stronger connections to place and higher learning about the history of the site.[21]

These applications of augmented reality to search functionality help us to rethink search in non-linear terms. We are used to linear models of search such as alphabetization and ascending or descending numerical cues, like you might find in the Dewey decimal system. Augmented reality invites us to rethink our search functions as spatial endeavors rather than linear ones.

Simulations as Augmented Reality

Third, augmented reality as simulation refers to the physical overlay of digital images, graphics, or projections in the experience of a physical environment. The human experience in cities faces challenges from urbanization that are being addressed by the design of sentient objects. As an example, in quickly growing cities, new micro-apartments are providing affordable alternatives for housing in tiny living spaces. As a result, windows and natural lighting may become a luxury in some densely populated regions, as trends in Hong Kong and Beijing are suggesting. One potential suggested solution for this disconnection from natural light is mixed reality living spaces. In a mixed reality living space, the physical and digital merge to create a hybrid space. In one hybrid concept,[22] designers suggest digital windows that are programmed to show various scenes, like streetscapes or natural scenes. These windows could utilize sensing technologies to respond to user needs based on furniture arrangement, behavior patterns, and time of day. The designers point out that using time of day as a component of the design is especially important as the circadian rhythms of daily life are fundamentally shaped by light.[23] A digital window might be proxemopetal if it displays a real-time view of the street in front of the building that houses it, or it might be proxemofugal while displaying a remote beach island. Like many technologies in this category, it allows for connection to or escape from place.

However, a digital window to the cityscape is only superficially prox-emopetal. When I began to look for examples of sentient objects creating proxemopetal relationships between people and spaces, I started to find them everywhere. In an Old Navy store, I encountered an interactive floor display that invited me to play soccer and chase butterflies. When I walked into Vanderbilt University's Jean and Alexander Heard Library, the WordsWorth display[24] caused gently undulating words projected on the floor to scatter as I walked through them. When I thought about the words running away from me on the floor, I wondered if I should read the movement as an homage to the thrill of the chase in research. Maybe this library was somehow protecting its words from me or even suggesting that the knowledge contained therein would always be out of reach. Or perhaps the puddle of words simply added levity and disruption to the concept of library. The fact that I paused to inquire, to think about research as play, and to experience the space was probably the most important criteria to measure here. The displays in Old Navy and Vanderbilt's library both used infrared cameras to sense my body heat and alter the visual projection accordingly. But they also made me pause and connect to the place.

At the Hafod-Morfa Copperworks heritage site in Swansea, Wales, researchers are exploring options beyond digital tour guides to connect visitors to the site and provoke a stakeholder discussion about how the heritage site can be best used.[25] Their options include two prototypes of performative technologies that invite visitors to interact with the physical space through digital media. The first prototype uses a pico projector connected to an iPod Touch, allowing the user to project information about the location onto the surfaces of the location. To encourage playfulness, the projectors are handheld so that users can project information wherever they wish, but users can augment reality by lining up the displays in various ways indicated on screen, fusing the physical and digital in space. In the second prototype, researchers experimented with spatial audio, allowing users to control audio using mobile devices. Multiple users working together from various distances can collectively modify the volume of the audio to localize or expand its reach. The researchers suggest that these performative technologies could be used alongside hidden markers to create a performative scavenger hunt, or they might be positioned in ways that disrupt normal pedestrian flow to encourage spectatorship when activated. This blending of space, technology, and performance capitalizes on the human behavior elements of proxemics and connects them to both device and information in creative ways.

Other technologies can be used to not only project onto spaces, but also build light displays in three dimensions in space. We commonly call these holograms.[26] For example, in Russia, the non-profit organization DisLife engineered an augmented reality publicity stunt to raise awareness about parking spaces for people with disabilities. DisLife reports that more than 30% of Russian drivers choose to occupy parking spaces marked for drivers with disabilities without concern. To combat this trend, the organization designed a stunt involving holograms projected into parking spaces. First, a camera senses whether or not an approaching car has a disabled identifier—a sticker—permitting the car to park there. If no sticker is present, a light illuminates, projecting the image of a person onto a water mist screen. The hologram, an individual in a wheelchair, says in Russian, "Stop! What are you doing? I'm not just a sign on the ground. Don't pretend that I don't exist."[27] The hologram later says, "Yes, I am real."

Play as Augmented Reality

Augmented reality as play refers to the use of physical elements in the real world to structure participation in a digital game. These might include digital scavenger hunts, contests requiring users to visit particular places and check-in on a mobile app, or games that inspire fun and creativity inside a particular physical space. Games can be structured to incorporate any number of the other augmented reality techniques or use mobile technologies to tie users to both a series of static locations and portable mobile technologies. The issue at hand is the one of play and how to incorporate technologies into play spaces, or remake spaces for play. At RMIT University in Melbourne, Australia, researchers Alan Chatham and Floyd Mueller added an interactive screen to a public basketball hoop to assess how interactive displays might enhance game spaces. The LED display was connected to a sensor on the hoop. At various times, it tallied the number of baskets made in one day, displayed the time between made baskets, and offered randomized written feedback (of the positive, negative, and trash talk varieties) to players. Users of the space reported that the display made them feel connected to the space and to the community of play surrounding that space. One player who often used the court alone at odd hours noted that a running total of baskets made that day let him know that he wasn't the only one around. To encourage the development of similar opportunities, researchers noted several design strategies for interactive game spaces, including (1) the display should be noticeable but not interrupt the activity of the game, (2) the display should motivate or drive the activity of the game, and (3) the display should share data between user groups to connect players to space.[28] These overarching strategies, which they describe in further detail, are fundamentally proxemopetal. They drive players toward the game, motivating them to play and enhancing their connection to place. This type of display helps us to think of future ways that augmented reality might create play spaces through connection to space via place-based technologies.

Navigation as Augmented Reality

Augmented reality as navigation refers to the use of an alternate routing structure to direct people through a space. Traversing routes using global positioning systems (GPS) and other interfaces builds an opportunity for augmented reality. GPS locates devices in geographic space and indicates user position on a map. Typically it is accompanied by another reading from the device's magnetometer, which acts as a compass, orienting the device in relation to the cardinal direction north. These data are typically overlaid on a two-dimensional map, a satellite image of the surrounding area, a photograph of the environment, or a live image from the device's camera. Although some people would argue that the latter example is the only augmented reality experience in the bunch, the positioning capabilities of our devices create an augmented reality when we use them to overlay our physical environment with their digital information. Even greater degrees of augmentation are possible if the device can project the map on the windshield of a car in motion, on the surface of a pair of eyeglasses, or onto the path or roadway being traversed.

Navigation, both in the context of augmented reality and in its many paper, mediated, and locative device–based forms, has emerged as one of the more researched constructs related to the human environment connection, and is of particular importance to the study of digital proxemics as a function of the ways we select movements and orient ourselves in spaces. So, let's spend more time exploring the influence of digital technologies and user-generated data on the practices of navigation.

Mapping (and Remapping) Our Environments

The work of mapping our geographic environments has been a source of contention since the beginning of modern cartography. Maps are sources of power derived from land ownership and regulation. The act of making a map and identifying physical structures gives legitimacy to a physical territory, and the plotting of deed lines, political boundaries, and territories confirms one particular reading of the map: that of its author. The seemingly simple act of naming sites on a map privileges one reading of that site's history. As an example of this contention, consider the multiple meanings of mountains. Cultural geographer Kevin Blake at Kansas State University has studied the narratives of place in the American west. In one study,[29] he examined six mountains sacred in the Navajo tradition, including one mountain in New Mexico noteworthy as part of a volcanic caldera with visible, hardened lava flows. The Navajo call this mountain Tsoodzil, or Tongue Mountain, while it is named on US maps as Taylor Mountain in honor of President Zachary Taylor. Taylor Mountain/ Tsoodzil, which also has additional names and meanings in Pueblo tradition, can be viewed from Interstate 40, which runs across its base. Even though its naming and land use have been contested, Blake notes that it is the least contested of the six Navajo sacred peaks. The most contested, San Francisco Peaks in Arizona, known by the Navajo as Dook'o'oslud or "Light shines from it," has inspired debates and lawsuits over mining, water, and logging rights at the site.[30] Maps, and the people that get to create them, have power over the spaces they reference.

In a theoretical critique of the practice of mapping, Jason Farman explored the digitization of maps as a sign of cultural change.[31] Building on cultural critiques of mapping authored by philosophers Jean Baudrillard and Guy DeBord, Farman reminds us, "Those who are in control of the design and distribution of maps are the ones in the position of power."[32] Digital technologies offer users the chance to control and distribute their own maps, upending the power held by traditional mapmakers. Digital maps, like those created by Google through Geographic Information Systems (GIS), offer users the ability to add data to maps, and creates the potential for public input into mapmaking.

In a fascinating example of this user-generated cartography, the popular podcast series 99% Invisible reported on the topic of Busta Rhymes Island, a tiny island in a community pond near Shrewsbury, Massachusetts.[33] A local Google Maps user, Kevin O'Brien, connected the moniker to the island through geotag applied in Google Maps. The name has not been officially recognized by the US Board on Geographic Names. O'Brien's application to name the island was denied because Busta Rhymes is alive (a person must be deceased for 5 years before a geographic feature can be named in his honor).[34]

Is the tiny island in a community pond in Shrewsbury, Massachusetts actually named Busta Rhymes Island? If you're Kevin O'Brien, it is. If you're viewing the island via Google Maps, it is. If you're a Shrewsbury, MA local, you've probably heard of it. If you're a representative of the US Board on Geographic Names, the island is not Busta Rhymes Island (yet). The answer depends on who has the power to decide a name. As Farman points out, and Busta Rhymes Island illustrates, user-generated maps have the ability to upend the power structures related to map authorship.

Farman, basing his argument in that of philosopher Guy DeBord, suggests that the creation of urban landscape is an exercise in power. The act of making streets, buildings, and intersections with traffic signals and entry points creates an environment full of scripts for specific types of interactions between people and their environments. People, Farman argues, have the ability to work within urban structures through improvisation, creating their own scripts of behavior in the moment[35] and even subverting the map as constructed.[36] Moreover, digital, user-generated information that overlays maps allows for much improvisation.

One example of this data overlay for improvisation was designed by the team at Strava Labs. Strava, a running and cycling GPS tracker, has produced a global heat map of cycling to indicate areas of cities where users most often cycle.[37] As of May 2014, the Strava heat map includes data from users generated by over 77 million cycling rides and almost 20 million runs, boasting over 220 billion data points. As long as Strava is in operation, that number will only continue to grow.

As one might imagine, the majority of our city maps are designed for automobile traffic. They depict where to turn to arrive at a destination, where to park, or the possible routes through the city. Other common city maps, at least in more pedestrian (or touristy) areas, depict city landmarks or promoted sites, guiding visitors to particular restaurants, hotels, and sights to be encountered.

On a Google Maps version[38] of Greenville, SC, the most important elements of the map include the interstate and highway systems. The Google map demonstrates traffic patterns in real time, thanks to mobile phone data. When studied side-by-side with a Strava heat map of Greenville, SC,[39] many prominent, geographic features of the two maps are the same, including Paris Mountain State Park, downtown Greenville, and Highways 25 and 276. The one noteworthy difference is the Strava map's boldest feature: a vibrant, thick road that does not appear on the Google Maps version of the same location. In fact, it is not a road at all, it is a 13-mile greenway developed in a rail-to-trail program in the city. Running north to south, the paved, multi-use greenway is a haven for runners and cyclers, but is closed to automobile and other motorized traffic. The overlay of the heat map on the geographic map demonstrates patterns of use in the city that cause it to be remapped, or viewed in a different way.

In an interesting study of the Strava interface, Roth Smith combined the study of technology and art to look at artistic improvisation of remapping.[40] Strava measures an athlete on multiple levels of performance including time, elevation, and geographic position. Noting that mobile fitness applications like Strava are changing the way cyclists choose routes, Smith investigated the rise of StravArt, a combination of Strava and art. Cyclists create StravArt by navigating the city in a pattern that, when overlaid on a map, creates an intended image. Around San Francisco, users are creating a variety of cycling routes ungoverned by typical transportation concerns, but rather artistic ones. The logo of the San Francisco Giants, a giant Thanksgiving turkey wearing a top hat, and the N'yan Kitten are among the rides that cyclists have drawn in the hills of San Francisco, California.

The ability to navigate a static urban environment in patterns of our choosing allows us to move through the environment with freedom and a sense of agency. In the case of StravArt, these choices to remap a city connect us to the environment and cause us to focus on our relationship to the site. The StravArtist is physically and temporally connected to his location, but also mindful of his movement. As with StravArt, the way that we use, add data to, and interact with maps dictates the experience we will have and controls the usefulness of user-data-based maps in our cities. The Waze app invites users to enter data about road conditions. Waze alerts connected drivers to road construction, accidents, and even hidden police officers. Data from Waze and other user-generated traffic apps are now being imported into navigation devices. Adding Waze's qualitative information to quantitative traffic flow data explains traffic congestion and informs users about their options in route selection.

The location of connected others can also influence our mapmaking. As Adriana de Sousa e Silva and Jordan Frith[41] note, locative social mobile networks mediate relationships not only between users and spaces but also among users who are all connected in a network. Building on the work of Jim Spohrer's Worldboard project,[42] de Sousa e Silva and Frith argue that data about place overlaid upon place have the potential to connect data and place in new ways. When social networks add location data to maps, we discover new reasons to visit (or avoid) certain places—our friends have been there, or are there currently. Such influence is not confined to urban areas and built spaces. In an interesting study of Kenyan pastoral herders, Bilal Butt examined the cultural contexts that underlie the impacts of mobile phone usage on clan-based herding patterns.[43] As herders encountered health issues in a herd, predators, or other location-based dangers, that information informed other herders in proximity. However, as Butt suggests, information sharing patterns and geographic movements among herders are related to the social, environmental, political, and economic conditions of the time at hand. The realized importance of nuance and context in this work reminds us that neither technology nor environment are necessarily deterministic in nature, so the ways we navigate through geography and terrain cannot be studied without an examination of the contexts for such movement. As social networks and locative media continue to evolve, the connections that relate people in networks to geographic maps in real time will continue to influence the ways we navigate our environments.

As we've seen with Busta Rhymes Island, StravArt, and the addition of user-generated data to street maps for real-time traffic overlays or geo-positioning of connected others, digital technologies empower users to contribute to digital mapmaking. When we make active choices about how to create maps and move in a city, we connect to the urban landscape, allowing digital layers of information to enhance our proxemopetal experiences.

The Dark Side of Directed Navigation

Augmented reality in its various forms, and particularly in the form of directed navigation, creates proxemopetal designs that facilitate connections between people and cities. These result from active mapping, interactive spaces that focus on user-experience design, opportunities for search and play, and augmented reality applications. But each of these proxemopetal interactions has its own dark side. Whereas these technologies have the ability to provide users with enhanced agency, seemingly proxemopetal digital technologies also have the capacity to become directive, removing agency from the user. One of the most telling examples of this occurs in street navigation.

Standing in direct opposition to enhanced agency afforded to digital media users in urban settings, digital street navigation may result in decreased agency for the typical user. The adoption of navigation devices and their integration into automobiles has fundamentally reshaped our experiences with wayfinding and navigation. Earlier, I defined wayfinding and navigation like this: wayfinding refers to directional cues and the resulting process for determining which way to go, and navigation refers to the act of moving from point to point. We typically navigate spaces in one of two ways: visual cues methods (looking for landmarks or using a map to develop a route) vs. stimulus-response methods (following turn-by-turn directions in sequence). In the first method, when we use visual cues for navigation, we are participating in the act of wayfinding by connecting with the environment and allowing it to aid in our navigation. When using a navigation device with turn-by-turn directions, we separate ourselves from the environment and defer our choices in wayfinding and navigation to the device.

My concern in the use of route planning and navigation devices is that they are built for us to use them by becoming passive. This passivity, evidenced in a lack of mindfulness, decreases our control of our own behaviors. Passive navigation-device use and reliance is certainly not proxemopetal use of digital technology, and it may even create a proxemofugal experience, separating us from the spaces we encounter. As we've seen, the technology doesn't concern me; our passivity does. Even though many navigation devices now use user-generated data for real-time route planning, the passivity with which we use them results in a technology-driven remapping of the city. I can recall a car ride to view a holiday light display during which the driver used turn-by-turn directional cues from a navigation device to direct us to the neighborhood. We completed no less than three (unnecessary) U-turns as instructed by the device, spinning the car in circles around virtual chicanes and medians that did not exist on the physical roadway. We navigated the car at the whim of the computer algorithm using an erroneous map.

This lack of mindfulness to location may even have interesting effects on our brains. In 2014, the Nobel Prize for achievement in physiology or medicine went to three researchers who identified the portion of the brain responsible for orientation in space.[44] Nobel winners John O'Keefe, May-Britt Moser, and Edvard I. Moser identified the neural connections between place and memory in the hippocampus and surrounding regions of the brain that encode routes and directions. Their collective work explains a series of recent findings[45] that navigation-device use might be detrimental to our long-term spatial reasoning.

When we become passive by allowing a device to dictate our movements, we may cause an unintentional disconnect between person and location. The navigation device user is physically and temporally present in the location, but is arguably less mindful of his connection to place. The map (and the map's designers) get to pick the best routes for us.[46] As driverless cars continue to be researched and developed, and ultimately released for public use,[47] they too will rely on these maps and varying levels of driver mindfulness. Their embedded maps and features will shape the ways we move in ways yet to be determined, but the creators of the maps will have the power. Maps determine our direction and orientation. They control our movements.

Although the descriptions in this chapter have referenced multiple technological interfaces and devices that create experiences for users, all of them require users to make a choice. As we've learned, digital technology on its own is neither proxemopetal nor proxemofugal—it neither connects us to a place nor isolates us from place. Instead, our choices about how we use digital technologies shape our interaction with place.

I'm not suggesting that we should go through life having only proxemopetal experiences, though being present where we are would typically be considered a good thing. Instead, I am suggesting that we should not go through life having mostly proxemofugal experiences, missing the presence of the physical environment around us. In urban settings, the busy-ness of life as a resident or tourist can seem overwhelming, causing us to rely on mobile devices to shape our experiences. As we continue interfacing with our cities digitally, my hope is that digital technology can connect us to our places with increasingly strong ties. In the next years, many decisions will be made by digital developers that will dramatically alter our connection to places. Their choices will shape the ways that place-based technologies either connect us to or draw us away from places. Moreover, they will fundamentally determine the boundaries of personal privacy, proximity-detection, and surveillance that we encounter in public spaces as the places in our lives and our technologies combine to locate us in space.

· 6 ·

LOCATING US

While perusing Portland-area maps in September 2013, Dustin Moore spotted something unexpected. Moore examined the street where his grandmother once resided. To look more closely at her house, he used Google Street view, a service that creates interactive maps through a series of images collected by cameras atop cars. As the cars drive up and down the road, their cameras archive visual images of the locations which are stitched together and uploaded to Google Maps. These images are updated by Google or its users at scheduled but varied intervals.[1] The result is a series of static images that capture physical locations at particular moments in time.

In Dustin Moore's case, the image on screen was shocking. Google Street View captured a panoramic photo of his deceased grandmother Alice sitting on her front porch before she died. In the photograph, Alice is shown sitting on her front steps happily reading the newspaper on a sunny day. "I thought it was such an uplifting and awesome picture because it showed just how laid back and awesome she was. Might not mean much to anyone else, I just thought I would share," Moore wrote on Reddit, an online content sharing community.[2] He says it might have been the last photo of her ever taken. Ultimately, Moore's Reddit post and the Google Street View image went viral and were featured on local, national, and international news broadcasts. After Reddit users across the globe reported seeing his story on news networks in the US,

Australia, and the UK, Moore quipped: "Grandma got a free trip around the world."[3] He replied to almost all of the reports and well wishes, thanking his fellow Reddit users for their thoughts and extolling his grandmother's enjoyment of life.

Over the past few decades, the intertwining of proxemics and digital communication has fundamentally altered the way we interact with locations. Stories like Moore's remind us that physical, geographic locations that have personal, historical, or communal importance are places that also bear digital inscriptions. To orient us to the vibrant conversation surrounding space and place, let me offer a brief synopsis here. In their compendium of *Key Thinkers on Space and Place*, Hubbard and Kitchin[4] articulate the ongoing conversation describing the distinctions between space and place as theoretical constructs. Space emerges from the disciplinary lines of geography, they argue. Originally characterized by a geometric relationship based on geographic location, space now incorporates both form and function of location as well as the social interactions that are inscribed on a space. Conversely, Hubbard and Kitchin characterize place as "relational and contingent, experienced and understood differently by different people; they are multiple, contested, fluid, and uncertain" (2011, p. 7). This articulation of place, based in a cultural studies model, suggests that places possess unique sets of characteristics that connect a visitor to the space of a geographic location, often in profound or unexpected ways. Places are therefore a subset of spaces—places exist in spaces, but not all spaces are places. The understanding of places and spaces as presented above is far more nuanced than the few words here suggest. Hubbard and Kitchin's introduction to *Key Thinkers on Space and Place* articulate the blurred boundaries of these concepts with an unusually precise and balanced examination of the issues involved, and readers new to the conversation surrounding space and place might find their reference text to be quite valuable.

The role of place is at once historical and connective, and the human use of space is fundamentally tied to our memory of place. For our purposes in this chapter, I'm interested in the role of digital technology as a creation facilitator, storage facility, and searchable database of information about space. In short, digital technologies offer new mechanisms for place-based archives.

An archive is a collection of primary source documents, and the term "archive" often refers to the physical location where these documents are housed. Archives are interesting to me because they generally serve a niche purpose. They are created by one user, one space, or for one purpose. For example, as a graduate student, I investigated political cartoons in the personal effects stored by Strom Thurmond, a potentially divisive historical figure, perennial South Carolina politician, and one-time US presidential candidate. In the Strom Thurmond archive, included documents were curated by Senator Thurmond throughout his life and are now stored in the archive to which he donated his personal correspondence. The political cartoons I studied were his personal collection of illustrations from his political career. As is the case with archives, copies of these same cartoons probably are stored in multiple archives in different types of collections. A cartoon of Thurmond from *The Washington Post* might simultaneously be stored in the newspaper's archive, in the Library of Congress, in a repository of political cartoons by the same illustrator, and in Thurmond's archive, four among many possible physical archives. Each archive is curated for a specific purpose, with a particular inclusion and exclusion criteria. Whereas Thurmond's archive has a set of political cartoons unique to that particular archive, each cartoon individually might also belong to other archives that include it. Digital technology offers us this same archiving potential. An Instagram photo of my mother's tomato pie might be stored in my parents' photo album, in my mobile phone's photo album, in my Instagram feed, and in searches of Instagram's #foodie hashtag, four among many possible digital archives. The primary difference between the physical archives I mentioned above and the digital archives noted below relates to access. Whereas I might have to travel to the physical archive to sit in a private room to view Thurmond's cartoons, the Instagram archive is mobile and quite arguably public.

The nature of archives as storehouses raises questions about how we access data in them. Archives hold volumes of data. These data must be studied and examined to translate data into information. Archives are accessible to select individuals and are curated by their owners. Thus, the issues of access and ownership, as well as translation, are wrapped up in digital technology and, as I will argue, tied to specific places in tangible ways. For Dustin Moore and his grandmother, Alice, Google Street View was an archive that connected person and environment in a historical context. As occurred for grandson and grandma, the archive has the power to shape the person-environment connection, not only in real time, but also across time and in relation to historical context. It is in the context of historical connections that places become significant to us.

Archiving Us

To better understand how digital technology shapes our connections to places, we must first consider how digital technology shapes the archives of space and place. Archiving a history of place is a three-tiered process when digital media are involved. First, we contribute information to the archive by purposefully adding our experiences to the data record. Second, our digital technologies contribute information to the archive by tracking and reporting our movements through overt and covert means. Third, the spaces we visit collect archives of our behaviors covertly as we move throughout the space. These three threads of data—us, digital technologies, and spaces—intertwine to form a bundle of information from which we can glean knowledge about places (and also about the people who inhabit them).

We Archive Us

The Newseum on Pennsylvania Avenue in Washington, DC has an inscription etched into its wall at the entryway: "Journalism is the first rough draft of history."[5] Attributed to *Washington Post* publisher Phillip Graham, the quote reflects the 20th-century focus on journalists as reporters and record keepers. The 21st century may see this idea in flux as journalism becomes the second rough draft of history. Social media has become history's first rough draft. Or maybe, social media has become journalism's first rough draft. Either way, we use social media to report on and keep records of our actions. It has become our primary archive.

And, as you might imagine, there is an app for that: Timehop. Founded in 2011 as 4SquareAnd7YearsAgo, the Timehop app[6] accesses posts from social media powerhouses Twitter, Facebook, Instagram, and Foursquare at yearly intervals, reminding us that the archive we created of ourselves is, for better or worse, both accessible and ever-present. A recent Timehop compiled from my own posts revealed a number of events that happened on the same date across several years: a five-year-old Foursquare entry when I took a new colleague and his family to lunch to welcome him to our campus, a three-year-old dinner party photo posted to Facebook, and a two-year-old Twitter conversation at a minor league baseball game. And, as Timehop reminded me, all of this happened on August 4th, the date in 1997 that Skynet was launched in the Terminator universe. Coincidence?

These social media archives are firmly tied to place. In March 2009, the launch of Foursquare signaled the rise of user-tagged geo-location as an archive of place. The Foursquare game asked users to check in at any and every location they visited. The game also rewarded players for checking in by assigning points for checking in, comparing point totals among friends, and awarding mayorships to the most frequent visitor to every location. In the summer of 2010, I served simultaneously (and proudly) as mayor of my university as well as the South Carolina–North Carolina state line, three produce farms in upstate South Carolina, two restaurants, and a rural Farmers' Market in low-country South Carolina at which we stopped on trips to and from the beach. In addition to mayorships, Foursquare also offered badges as rewards to frequent users. Cleverly titled badges like "Gym Rat," "Super Mayor," "Swarm," "Bender," and "On a Boat" made checking in surprisingly fun and motivated players to seek out opportunities to receive one of these awards.

Early users of Foursquare added geo tags to locations, enabling other users to search and find locations nearby. Businesses, particularly those in urban areas, crafted financial rewards for users who checked in while visiting (I remember choosing one coffee shop over another because of the 10% discount it offered if I checked in on Foursquare). Foursquare users could add photos and reviews of their experiences as a personal archive of spaces visited, and as a crowd-sourced archive of experiences at a location. At my university, students in one of my social media classes completed a class project mapping our entire campus on Foursquare. They quickly became the mayors of every fountain, hangout, building, and classroom while archiving their movements on campus.

In March 2010, Twitter capitalized on Foursquare's emerging popularity by adding location tags to tweets. Facebook followed suit that August.[7] At the time, many concerns related to location sites—such as decrease in privacy, invitation of crime, and opportunity to run afoul of stalkers—were raised by social media users. Dan Fletcher's *Techland* report for *Time Magazine* in 2010 questioned the role of location-based applications like Foursquare in daily life: "It's a bit too much like having a pint-size version of my mother in my pocket, constantly prodding me for updates. Where are you going? Who are you with? What are you doing?"[8]

Despite these early concerns, people have readily adopted location-based check-ins in moderation. By 2013, the Pew Research Center reported two interesting findings related to location-based services.[9] On one question, 30% of adult social media users reported that they had at least one account with geo-tagging enabled to include location in posts. Conversely, in another question, 48% of teenage users and 35% of adults had turned off location-tracking on their phones. This doesn't mean that they weren't using check-ins sporadically, but rather that they knew how to turn off their location services if desired. These two findings together indicate both a growth in the use of location-based services on mobile devices and some awareness of personal choice in the creation of one's own archive.

But geo-location is not the only means of creating a user-generated archive. Irrespective of our choice to geo-tag our posts, we create an archive of place with every tweet, status update, photo, and video we share. I might reveal my age or my penchant for photography with this next statement, but I love photo albums. The photographic archive of a physical album is a flipbook of my history and that of my family. At our house, we store them in the family room and fill them with the visual records of our happenings. They are private artifacts. As "photo album" becomes less about a book and more about one function of the mobile device, the archive is no less private as it exists on the device. However, as soon as we share a photograph through social media, it becomes part of our collective public archive.

In their book *Mobile Interfaces in Public Spaces*, media theorists Adriana de Sousa e Silva and Jordan Frith examine the role of digital information uploads in relation to urban and public settings.[10] They suggest that the organization of public spaces itself will change as people continue to add information and data to shared physical locations. Public spaces, once organized according to geographic proximity or event types, might be reorganized according to popularity, interface potential, networking capabilities, and the networks that participate in and with the space. The ways that we socialize in these spaces, they argue, will shift with the technologies that locate us. The archives we create will in turn reconfigure the spaces we inhabit.

Not only do these archives we create on social and mobile media reconfigure spaces, but they also have the capacity to alter our memories. Writing for the *New York Times*, Teddy Wayne commented on the archive created by digital photography shared online. In an interview with Dr. Linda Henkel of Fairfield University, the two explore the impacts of photographs on memory.[11] As we create our memories, we have the original experience, which we translate into a story for retelling, with various layers of exaggeration. When we take photographs, the photo doesn't change over time the same way that the story does. Soon, we lose the story and remember only the photograph, judging it as more accurate. In addition, the filters and other photo editing tools applied to photographs distort and add bias to our memories. An Instagram filter makes a day look sunnier than it was. A retouching tool can remove that pesky coffee stain on your shirt or a blemish on my cheek. Memory is changed. Moreover, our memories are reshaped by the captions we write about our events alongside the likes or comments appended to our shared photos. A photograph liked by two hundred of my friends might become a more valued memory than a photograph liked by only a handful of people, simply through the positive reinforcement provided by social media approval.

Scoffing at social media's role in archiving memories, columnist Joel Stein titled a column on the subject as follows: "Don't speak, memory. If the information age doesn't shut up soon, all our best stories will be ruined."[12] Actual memories have a way of being solidified in emotions rather than in fact. The archive we create of us reminds us of both—emotions and facts—in a compilation of moments tied to places.

In the context of social media, this archive we create becomes a creative process of identity management. Sociologist Erving Goffman famously discussed impression management and our choices about what, when, and where to share information. In his essay on footing,[13] he discussed the people, the co-present others, who are permitted to become an audience of our identities. In the physical world, these include the people we intend to address, as well as bystanders and eavesdroppers. As media researcher danah boyd has argued, these audiences and contexts are collapsed in social media.[14] Our management struggle becomes designing information meant for multiple audiences simultaneously. Furthermore, we also collapse the context of space and place. At once, our archive appears where we are and where our audiences are. Social media posts have the double effect of taking our information to our audiences' locations and bringing our audiences to our location. When I post a photo of my family sightseeing in Puerto Rico, my audiences get to come with us, even if just for a moment. Likewise my place-based photo transcends place to arrive around the world. This is exactly the same concept Dustin Moore wrote about on Reddit when he said, "Grandma got a free trip around the world."[15] She went around the world, but at the same time "around the world" came to her in Portland, Oregon. The archives we create are inextricably tied to place, and digital archives we create connect us to places as we create and consume them. But we are not the only creators of place-based archives.

Digital Technologies Archive Us

When we walk, we leave behind footprints that can be followed and interpreted. When we move through built spaces and urban centers, we also leave digital footprints, thanks in large part to the technology we carry with us. The most obvious of these imprints are a result of our carrying and using mobile phone technologies which locate us geographically. However, the digital footprints left behind by users are attached to a wide variety of things on the mobile device, including app usage, digital media capture, Wi-Fi connections, and geo-location. The digital footprints we leave in spaces teach our technologies about our behavior, and they report data, both overtly and covertly, to a number of archives.

Data tracking using mobile technologies creates opportunities for interactions between people and spaces, mediated by the sensing capabilities of mobile devices. Typical smartphones contain various sensors that can collect data to enhance a user's experience. These include several features that geo-locate devices, such as global positioning systems (GPS) and cell identifiers (GSM, location area codes, cell IDs), as well as low spatial coverage systems like Wi-Fi and Bluetooth technologies. In addition to location sensors, smartphones also contain other types of sensing technologies. These include the obvious microphone and camera, but also more subtle sensing tools like accelerometers, which measure speed of motion; magnetometers, which orient our phones in space; and light sensors, which auto-adjust brightness of on-screen display. These sensors can all collect and report data with and without user intervention.

For example, mapping applications that detect traffic patterns overlay them on maps to help users discern routes that minimize travel times. Current models of traffic condition detection rely on dedicated probes on major roads or GPS data from mobile phones. Computer scientists at Zhejiang University in Hangzhou, China, are designing a digital system to detect traffic conditions with cell phone accelerometers.[16] The accelerometer, they argue, allows for a data collection system that is unobtrusive, adaptive, and energy-efficient. This means that it wouldn't rely on an app or user reports. Instead, the mobile device would automatically report congestion events by learning the patterns of vehicular motion, and would operate separately from the GPS-system, which consumes large volumes of power. The model is an information sharing concept that tracks and stores users' digital footprints to serve the greater population. Likewise, Designers in the Akashi National College of Technology in Hyogo, Japan, have proposed using smartphone light sensors to assess lighting in pedestrian areas at night.[17] Their concept is to infer data about spatial street illumination from the movement of the light sensor across time through areas of light and dark. As the light sensor notes a transition from light to dark, that data paired with geo-locative data can create a map of lighting patterns. Such information might be used to report pedestrian areas with sub-standard lighting to city work crews or even alert pedestrians to avoid streetscapes with minimal lighting. In a third example, a group of researchers at Yonsei University in Seoul, Korea, have developed an algorithm that tracks pedestrians in indoor corridors. Using a magnetometer (which powers a cell-phone's compass) and the phone's accelerometer, the algorithm can track where users are in corridors, their speed of travel, and their navigation of corners and other obstacles.[18] The algorithm accurately described pedestrian behaviors even under conditions where the phone was in use or being moved, swung, or transferred among multiple uses.

The examples above all indicate ways that information can be gleaned from mobile devices in transit with the user. But mobile phones can also pair with data collectors embedded in spaces using wireless and Bluetooth technologies. One quickly expanding example of the pairing of sensing technologies and physical space that we haven't yet discussed is geofencing, the use of fixed digital perimeters to trigger actions on mobile devices, such as sending alerts or notifications, unlocking features on an application, or blocking use of particular applications. According to Sandra Rodriguez Garzon and Bersant Deva in the Telekom Innovation Laboratories at Technischen Universität Berlin,[19] geofences can be built using geometric or symbolic designs. Think of a geometric geofence as a circle on a map. When you move into the digital circle, you are inside the geofence. As an example, I participated in a "Park Hop" contest that utilized geofences to unlock clues when I arrived within a mile radius of each park featured in the event. Conversely, a symbolic geofence would use hierarchy-based descriptors or wireless access points to describe a location. When you connected to the wireless routers or cell towers in the correct location, you would be inside the geofence. For example, in an office building, certain floors could have geofences that opened resources for people on that floor, but blocked access for others. Combinations of geometric and symbolic geofences, theorized as semantic geofences, could then be employed as well, allowing for advanced geofence adoptions.[20] In this scenario, geofences could be layered on top of and within each other so you might have to be present in a geometric geofence before unlocking a symbolic one.

The symbolic geofences described above relate to proximity detection—how close a user is to a location in question. One inexpensive hardware solution for the symbolic geofence is the beacon. The iBeacon, released by Apple in 2013, inspired a wave of iBeacon-compatible devices that function in conjunction with Bluetooth technology on mobile devices. In contrast to the geofence, the beacon does not use GPS to map latitude and longitude. Instead, the beacon recognizes devices in range of its signal, which is usually around 50 meters. It detects the user's proximity to itself. When placed in a location, it operates similarly to a geofence but using a different technology and, notably, does not rely on the battery and data needed for GPS connection through the device.[21] Because beacons do not utilize GPS, this means that, in addition to their placement on stationary geographic locations, they can also be located in moving locations, like buses, cars, or trains.

Geofencing and beacon technologies have myriad applications for business, gaming, wayfinding, and artistic applications and everything in between. They connect spaces to information through physical access points by connecting with the portable mobile devices that move in and out of proximity to it. We like to think of the Internet as a provider of information that travels everywhere for anyone. Proximity-detection technologies have the ability to contain information in physical sites to provide selected information to anyone who travels to find it, creating spaces embedded with technology that binds information to location, and for specified users.

Digital technology's contribution to this era of big data is not only in the acts of capturing data and sensing other technologies in proximity, but also in the work of translating data into usable information. In one example of this, Mark43 is a start-up dedicated to compiling police records in the cloud to enable all police offices to connect records in a massive searchable database.[22] Mark43 functions like a social network of police reports, connecting relevant information about suspects across precincts in real time. Software of this kind translates the vast data into relevant and user-friendly information.

Researchers at Wireless Sensor Networks Lab at Technische Universität Darmstadt in Germany have designed a technology called Locator which uses smartphone data to recognize places.[23] Locator combines user data, geo-location data, and algorithms to recognize the relevant places visited by the user, judging places as more or less relevant. Not only does this type of technology pinpoint the user's location, but also intelligently infers the importance of that location to the user. The applications of data that represent the *values* of a user, as Locator aims to provide, change the nature of data collection. No longer would smartphones collect user data, they would also be able to translate that data into information and knowledge about the user. Such knowledge could be used to compel us to visit, donate resources, contribute to causes, suggest related locations, or be sold to advertisers and others with additional motives.

Researchers at the University of Cambridge explored communication patterns in one organization as the company moved from one building to another, allowing them to compare the influence of built space on organizational communication.[24] The team used RFID (radio frequency identification) badges to capture face-to-face interactions between people in the old building and the new one, allowing them to compare the differences in use of space. Their study lends support to the notion that shared informal spaces create greater opportunities for interaction in the organization. Furthermore, its use of sensing technology to measure employee interactions provides a new mechanism for researching interpersonal use of space, provided that research subjects consent to being tracked.

This organizational study using RFID chips indicates a subtle change in data tracking. Data tracking produces volumes of data that can be synthesized into information. And computers are developing the ability to not only track behavioral data in spaces, but also to infer the human activities occurring therein. Just as online search engine algorithms infer our interests to provide recommendations, digital technologies in spaces are learning how to use algorithms to translate our movements into inferred behaviors. In the translation of data to information to knowledge, digital technologies tracking us become physical spaces watching us. It is only a tiny leap from there to physical spaces being able to alter themselves based on our behaviors to meet our needs.

Spaces Archive Us

Separate from and connected to user-generated archives and traces left in and by digital technologies, the spaces we inhabit and visit are also archiving our behaviors. Most people would refer to the creation of this archive as surveillance. Surveillance is one space's attempt to capture and interpret the actions of a variety of users who move within that space or geographic location. Archives in spaces are no new thing. Using guest registries, audio recorders, or video cameras in businesses, on city streets, and in homes have created vast archives for any space that uses them. These archives are full of data that rest until needed, requiring humans to come in and perform the work of digging through archive footage to mine the data for helpful information.

Digital technologies require us to examine the negotiation between people in spaces that are no longer static, but, by the design of embedded digital technology, are active, dynamic, and responsive.[25] When connected to spaces, this type of software can be used to infer, map, and direct human behavior. One current emphasis in the study of our interaction with spaces relates to the ability of the responsive space to capture, or sense,[26] our use of the physical space. In *Pervasive Computing*, Nicolai Marquardt and Saul Greenberg[27] described their work creating relationships between ubicomp and proxemics. Specifically, Marquardt and Greenberg discuss the ways that mobile devices might be able to ascertain five elements of proxemics: distance (measurable area between people or people and objects), orientation (which direction a person is facing), identity (exact identification or ability to discriminate between people/objects), movement (changes in distance or orientation over time), and location. These five concepts—distance, orientation, identity, movement, and location—are constructs that we assess naturally in proxemic interactions. However, digital devices with the ability to make the same assessments can offer a wealth of data for information collection and interaction potential.

When considering the ways that technologies might sense user interactions and connections, Marquardt and Greenberg ultimately suggest, "People shouldn't act on technology but rather through technology to perform the task at hand."[28] To that end, the authors offer suggestions for ubicomp related to six challenges of personal computing: (1) revealing interaction possibilities; (2) directing actions; (3) establishing connections between devices; (4) providing feedback; (5) preventing and correcting mistakes; and (6) managing privacy and security. These are all areas in which digital technologies could serve user experience in a space by sensing and inferring the behaviors that are occurring within.

Examples of this inferring capability are increasingly common, particularly among ubicomp designers. In Italy, a team of researchers used mobile phone data to collect information about the interpersonal distance between people and their body orientation.[29] Having these two measures allowed accurate predictions of whether or not a social interaction was occurring and whether it was a formal or informal interaction. The predictive value of these two data points was so precise that the position of the phones on users' bodies made no difference. This research suggests that unobtrusive data collection from the position of mobile phones would allow similar software in spaces to infer social interactions in the space. Similarly, a second research team in Italy developed an algorithm that infers behaviors about social interactions, this time based upon images of free-standing conversation groups.[30] Consider a cocktail party for example. Groups of people form and disperse in space, focusing on conversation, and then break apart and regroup. When captured from above, images of these group patterns share particular shapes such as circles, horseshoes, and L-shapes. These positioning patterns are collectively termed facing formations or F-formations. Images, like unobtrusive cell phone monitoring, provide enough data to accurately predict social interactions in groups. If the two of these data collection points were combined, vast data about groups gathering in spaces could be translated into an archive of encounters, connections, and personal affinity, all without a human needing to watch the footage.

One group of researchers has proposed a model to aggregate data in networked communities to reveal patterns in community behaviors.[31] Noting that activity recognition data sets typically focus on the individual level, they propose that capturing collective behaviors in communities might allow for more robust data translation. Robustness increases the likelihood that the software will accurately infer the recognized behavior. In one example, data measured by a workplace on a personal scale would depict wide varieties of commute times and locations. However, when viewed collectively, the commute times and location displays would all sync in arrival time at the workplace, allowing the software in the workplace to infer the act of commuting accurately.

These types of sensing and inferring devices, when placed in spaces, enable spaces to become sentient. Smart rooms, smart buildings, and smart public spaces capture data, translate it into information, and store it in a place-based archive separate from the archive of the user and the archive of the device. They can use both of the other archives, but often without user knowledge or control.

The archived history of a geographic location is a funny thing. It is often fragmented, hidden to some and revealed to others. For example, take the geographic location of my home. I know who owned it immediately before me, but beyond that knowledge, I would have to do a great deal of research to understand the events that transpired on my quarter-acre parcel of land over the past several centuries. Presumably (as I live on the east coast of the US) this land was once a piece of a British colony, and perhaps previously inhabited by Native Americans. Beyond that generic knowledge, I may never know about this geographically located piece of land. Another visitor might, however, give more knowledge of the site. A friend recently relayed to me the story of taking her husband to see her childhood home in a different state. As they arrived at a 1950s house in Union Township, New Jersey, she delighted in seeing her home as she remembered it. The two pulled up to the house, took pictures on the sidewalk, and she relayed stories of her memories of the home. And the current owners came out to say hello (and probably also to investigate why an unknown couple was taking photographs of their home).

This geographic location had a particular history for her, and she shared that history with the people making their own new histories in that place.

These histories of place are stored up in memory. And these memories often contradict one another. One version of memory of a geographic site is chronicled in a historical lineage of deeds and purchases. Another version is captured in the minds and emotions of past visitors. Still other versions of memory of place are captured in stories told and retold from generation to generation. Digital technology threatens to alter human internal mechanisms for storing memories of place. Digital records capture us without emotion, ambivalent to context, as we are in space. The recordings captured in image and video tell a story of place from the point of view of an omniscient third party—a digital narrator in our midst. In the *New York Times* column discussed earlier, Teddy Wayne suggested, "When we 'redo' a memory by playing it back, the recreation even surpasses the original."[32] In contrast to memories that we store in our brains, surveillance footage documented through video, audio, and text can be slowed, read, re-read, zoomed in upon, and scanned through a variety of applications with the potential for lip-reading, facial recognition, and external interpretation.

This type of surveillance is common, frequent, and pervasive, particularly in urban centers and public spaces, in addition to guarded private locations. As the history of surveillance continues to teach us, every advancement in technologies of surveillance are soon accompanied by technologies that counter the effects of surveillance. In a study of resistance to surveillance, Torin Monahan explored anti-surveillance camouflage as a response to pervasive surveillance in public places.[33] He describes examples of anti-surveillance camouflage including hairstyles, makeup techniques, and face masks that disrupt facial recognition patterns, clothing that hides thermal emissions to escape detection by drones, and fashion that contains reflectors or emits infrared light to distort photo and video tracking. "Anti-surveillance camouflage of this sort flaunts the system, ostensibly allowing wearers to hide in plain sight—neither acquiescing to surveillance mandates nor becoming reclusive under their withering gaze."[34] His argument suggests that aesthetic responses to surveillance, like camouflage, fundamentally accept surveillance as a normal condition of life.

Maybe our collective culture will not be bothered by pervasive surveillance through sensing and inferring technologies. We might just accept surveillance as a part of everyday life, as Monahan suggests might be indicated by the creation of anti-surveillance camouflage. But, whether we accept it or not, our movements will be changed by pervasive surveillance as society collectively considers the impacts of being watched, and archived, as we move in space.

On Privacy and the Creation of Archives

The three modes of data collection based on location share a complicated relationship with personal privacy that is worthy of a brief mention and pause here for reflection. In a 2015 report, the Pew Research Center explored Americans' views on privacy, security, and surveillance.[35] The report argues that the majority of Americans value the ability to maintain privacy and confidentiality in both online and offline settings. One survey discussed in the report asked questions about information collected when we navigate online spaces, information collected as we move around in physical public spaces, and information collected at work or at home. The report suggests that two major factors influence our views on privacy: permission and publicness. Some of the findings speak to these factors: 88% of Americans found it important that they not be watched or listened to without their permission; 63% of Americans found it important that they could "go around in public without always being identified." And, whereas 85% of Americans placed importance on not being disturbed at home, only 56% placed importance on not being monitored at work.

In the 1990s, Internet users enjoyed relative privacy while browsing online. Public awareness of the slow erosion of that privacy was not widely amplified until 2013 when Edward Snowden alleged that the National Security Agency (NSA) was monitoring Internet users' browsing histories. The US government, along with corporations and service providers, had been tracking Internet users for quite some time, indicating that the Internet stopped being private sometime around the year 2000. Inger Stole, a professor at the University of Illinois at Urbana–Champaign, credits corporate America for that change. "Contrary to common belief," she writes, "the driving force behind the Internet's transition from a place where users were anonymous and in charge to one where individuals were constantly monitored by powerful unaccountable interests was not national security, it was commercialism."[36] Businesses trying to sell their products invested in market research through online monitoring. As we allowed "cookies" and entered our own data, corporations bought and sold email addresses, IP addresses, and data about user clicks. This data harvesting has afforded Internet search providers the ability to develop algorithms for enhancing search for maximum user experience.

In his book *The Filter Bubble: What the Internet Is Hiding from You*, Eli Pariser writes about the alteration of search to focus on the user.[37] In contrast to the original concept of search engines which might return a list of links based on their relation to the search term, a current search engine uses the searcher's browsing history and other personal characteristics to infer what links the user might enjoy, and only reports those. As you might expect, two people sitting together in the same room on their own devices would thereby receive different results for the same search term.

Reporting for *Time Magazine* in 2010, Lev Grossman examined search engines as recommendation engines, the concept that a search engine could be used to tell you what you might also like.[38] Online retailers, most notably Amazon.com, have used this concept to compel users to discover and ultimately buy additional products. As Grossman points out, the development of recommendation engines was not without incident. In one of the more disastrous downfalls, a search for a movie at Walmart's online site in 2006 resulted in a backlash over race relations against the organization. When a user searched for *Planet of the Apes*, the website also suggested DVDs about Martin Luther King, Jr., Dorothy Dandrige, and *Unforgivable Blackness: The Rise and Fall of Jack Johnson*. Walmart responded quickly by altering the algorithm and issuing a media release: "Walmart.com's item mapping process does not work correctly and at this point is mapping seemingly random combinations of titles. We were horrified to discover that some hurtful and offensive combinations are being mapped together."[39] Walmart also noted that the "bizarre nature of this technical issue" was also returning African American literature alongside searches for *Home Alone* and *Power Puff Girls*. Recommendation engines have gotten much more precise in the decade since the Walmart incident, but they remain more art than science, as many are designed to use user ratings, reviews, and viewing patterns to make suggestions for other users. As we discussed in chapter one, Pandora, Amazon, Netflix, and Google—among many others—use these algorithms to enhance our online search experience.

When paired with location-based services, Internet monitoring creates localized, directive recommendations for the user. Even though geo-locative apps began as a way for users to input their locations, many now help to direct user movement. As journalist Dan Fletcher reminds us, "The ultimate aim of geocentric sites is less to pinpoint your location than to offer guidance on what experiences you should seek out (or avoid) there."[40] Groupon, Living

Social, and other online promotion sites pair algorithms for search with location-based service to signal "deals near you." When I'm searching for a hardware store, Google uses my location to filter responses so that I'm directed to
the closest store near my location, not the most searched-for store in Detroit.

Because they are overt and self-generated, user-generated archives seem
most attuned to balancing permission and publicness. We choose to record
video, audio, or photographs of the events in our lives, and we can keep those
recordings private if we wish. However, our digital data collection technologies are designed for sharing. When I open a video recorded on my iPhone,
the following options appear: AirDrop, Message, Mail, YouTube, Facebook,
Vimeo, Vine, Instagram. Scrapbooks are no longer a private set of home videos on dusty VHS tapes and collections of photo books. Digital scrapbooks
are shared by nature of their design: blogs, Facebook walls, YouTube channels,
and Instagram feeds. In these cases, we make conscious decisions about what
and how to share. These acts of sharing balance the concerns of permission
and publicness discussed above as we consciously and purposefully share our
own information, to our benefit or detriment. Researcher Alice Marwick[41]
explored the concept of social surveillance and its positive effects. Marwick
distinguishes social surveillance from surveillance as an information gathering practice requiring mutual power in social relationships, lateral hierarchy
of individuals, and reciprocity of data sharing. The act of creating our own
archives online in conjunction with those of others, she argues, affords us the
opportunity to monitor and be monitored by each other. Because we choose
to share these archives online, we also feel empowered to peruse the archives
of others. This mutual sharing has positive effects ranging from increased feelings of digital intimacy, greater levels of collaboration, higher social status,
stronger relational ties, and enhanced social capital.

But, sometimes, we use our social archives to perform acts of surveillance
of others by recording unusual things happening in the world around us.
These recordings are often of other people, and we expand our archives to
include them or contribute their likenesses to crowdsourced digital archives
of place, tagging the location to prove we were there. Whether our sharing of
recordings of other people's lives respects their views on the balance between
permission and publicness is best judged on a case-by-case basis. In many
private settings, like parties in people's homes and family events involving
children, many people commonly ask permission to share photos or videos
before taking them. In public places such as parks, shopping centers, and community events, people seem to be more apt to share at will. One of the most

unusual place-based digital archives remains People of Walmart,[42] a website driven by user uploads of people shopping at Walmart stores around the US. The photos are searchable by date, location, user rating, and descriptive tags and are shared by the site's social media feeds. After looking at some of the pictures, I have asked myself whether some of these folks are trying to make an appearance on the site.

The 2015 Pew report on privacy, security, and surveillance[43] also reveals that 93% of Americans place importance on controlling who can access information about them and 90% place importance on controlling the types of information accessed. These data demonstrate a desire for control of personal data. However, only 9% of Americans felt that they had substantial control over who collects data or what data are collected. The disparity between wanting control and actually having it is striking.

In another survey discussed in the report, respondents were asked about mechanisms for maintaining control of personal information, such as clearing cookies or browser history, refusing to provide information irrelevant to a transaction, providing false information, using a temporary username, encrypting phone calls or messages, or using an anonymous web service like a proxy server or virtual personal network. Whereas almost 6 in 10 Americans had cleared their browser histories, only 1 in 10 reported using advanced techniques for enhancing privacy, like phone encryption, private networks, or proxy servers.[44] Regrettably, they did not inquire about the use of anti-surveillance camouflage.

These findings speak directly to the issues inherent with digital behavior tracking and locative media.[45] We leave markers everywhere we travel online, and the sites we visit leave markers on our devices. These markers transmit data to a variety of Internet sites. On the issue of who is collecting data, we know many of their names. Among those collecting information are advertisers, search engines, social media and video sites, Internet service providers, credit card companies, and government agencies. What specific data they are collecting is another question. Some of these sites try to address the issue of permission by having us read the terms of service and accept user agreements. However, others of these organizations operate behind the scenes, tracking user data without explicit permission. Still, our knowledge about the types of data collected is so lacking that many privacy experts warn that we should assume that everything we do online occurs in relatively public space.

If everything on the Internet can be argued as public space, then we have truly entered the domain of ubiquitous digital surveillance—I defined it earlier as one space's attempt to capture and interpret the actions of a variety of users who move within that space or geographic location. So far, I've discussed surveillance in relation to public and private built spaces. If privacy experts suggest that online spaces are public by design, then all of our actions in online spaces become subjects of surveillance.

Earlier, based on the Pew Report on privacy, security, and surveillance, I wrote that permission and publicness were the two major factors describing Americans' attitudes about privacy. Unfortunately, we all disagree about how these factors operate online. Permission—the act of approval about what information is shared with whom—continues to be a contested issue in digital media. Data collectors might argue that by using the Internet, we are giving sometimes explicit, sometimes implicit, permission for data collection through both overt and covert means. Conversely, users might argue that the only data that should be collected are the pieces of information for which permission is asked and granted. Still others, like some governments, argue that permission is an unnecessary consideration when the interests of national security are involved.

Publicness—broad access to space or location of an event, and the Relative opposite of privacy—is also a contested construct in online spaces. Data collectors might argue that the Internet is a public environment in which we can each choose our actions. In addition, the designs of services on the Internet are increasingly adaptive, allowing the online spaces we encounter to flex based on our choices. In this way, data collection is a tool used to create a better user experience. Users might argue that the devices and services we use give us the illusion of, if not also the expectation of, privacy. The US Supreme Court bolstered this claim with its 2014 decision in *Riley v. California*,[46] which required warrants for searches of cell phone data. Although legal scholars continue to debate the specifics of the decision and its actual impacts on privacy, privacy supporters heralded this decision as a protection of Americans' rights under the Fourth Amendment, which prohibits unreasonable searches and seizures.[47] The court suggested that the mobile device's ability to store immense amounts of information could reveal volumes of data about a person's history, noting the storage of data related to web search, application use, and geo-location. If the device contains private information, are the data collected and stored by data collectors also private? These questions will arise repeatedly in legal cases related to privacy. But the legal definitions of privacy and our expectations of privacy may actually be two different things.

In *Smart Mobs*, Howard Rheingold offers important, well-stated, and increasingly timely questions about pervasive surveillance: "Who snoops whom?

Who has a right to know that information? Who controls the technology and its uses—the user, the government, the manufacturer, the telephone company? What kind of people will we become when we use the technology?"[48] These questions cause us to reflect on how these archives are accessed, by whom, and for what purpose. An archive is a repository for historical data on a particular topic. Although we may control our own personal archives, digital technology both beckons us to contribute to archives of place, and contributes to archives of place on our behalf. Moreover, our movements in place are subjects of observation and now inference in the spaces we visit. We are moving archives that are building archives and being collected by archives. All these data can and will be translated into information for various purposes. For me and other researchers of proxemics, these data can give us a look at our culture in aggregate. For owners of spaces, these data can be used to enhance user experience. For still others, these data can provide a window into our preferences in increasingly great detail. The ways data will be used will continue to inspire heated debate, and as our society grapples with these archives, our collective notion of privacy and our cultural understandings of proxemics hang in the balance.

· 7 ·

INHABITING NEW ENVIRONMENTS

When I was in second grade, we celebrated career day. In addition to the parade of visitors sharing their lessons learned in the quest for vocation, every student in class wore the uniform of his or her planned career. My classroom was filled with doctors and lawyers and a few teachers. I wore khaki cargo pants and a shirt with a giraffe on it, owing to my dream of becoming a zoo-keeper. As you can probably guess, my seven-year-old self would likely be a little disappointed right now. But, at age seven who really knows what they want to be when they grow up?

At age twelve, I went on a behind-the-scenes tour of an elephant enclo-sure with a zookeeper. I marveled at the elephants and learned about their diet and daily routines. At age sixteen, I took a job at a veterinary clinic that served as the healthcare provider for a local zoo. On my first day of work, we had a baby lion in the clinic. They're much cuddlier when they are little. It was a common occurrence to find sea lions in the bathtub, owls in cat carriers, and peacocks on parade at work. But still, I was no zookeeper.

EarthCam changed all of that. Long after I decided to pursue a career in academia, EarthCam arrived at the Greenville Zoo. EarthCam is an interna-tional network of live streaming cameras that watch over public places, con-struction sites, and historically significant venues among many other places.

One contributor to our pervasive surveillance among many, EarthCam's tagline is "Where the world watches the world."[1] EarthCam has cameras all over the world, but anyone can see the streaming video they produce in real time. One division of EarthCam is connected with zoos. On October 22, 2012, over two hundred thousand people accessed a webcam installed in the Masai giraffe enclosure at the Greenville Zoo. The audience, myself included, watched as Autumn the giraffe successfully gave birth to baby Kiko at 11:49 p.m. After all these years, digital technology made me a zookeeper, at least for a moment, by allowing me into a place that only zookeepers get to go, at a time and in a place inaccessible to visitors.

Digital technology can offer us access to spaces that we might never otherwise experience, making the forbidden now available. Digital cameras can transport us to places too small, too remote, too dangerous, or too private to encounter in the physical world. Internet-based video sharing can move audiences into our space and project our spaces to locations. And livestreaming connects us not only in space but also in time. Furthermore, virtual reality technologies are beginning to offer immersive digital experiences that create a new genre of spatial experiences that we are only beginning to imagine.

Edward Hall described features of space as fixed, semi-fixed, and dynamic.[2] Fixed features are those real or imagined artifacts or boundaries that are perceived as immovable. Semi-fixed features are real or imagined artifacts or boundaries that are perceived as moveable but possess norms that often leave them in place. Dynamic features are typically moveable and perceived as inherently flexible artifacts or boundaries in space. The characteristics of features in a space shape the ways that we move in them. When digital video capture devices become features of the physical space, the space itself is changed as are the relationships between the remaining fixtures. Even when I hold a video conference, I adjust my camera to include and exclude selected items (and co-present others) in my physical space. So, when I consider digital technology's impact on the features of space, I immediately start pondering the opportunities for and limits of viewing the physical world through a digital lens. At some fundamental level, at least with current technology, the sensations of the physical world merge into two senses: sight and sound. As Hall noted, proxemics is bound up in the fullness of all the senses. Digital technology collapses the senses, and that action impacts proxemics in some obvious ways and in many ways we have not yet realized.

When viewing the physical world through a digital lens, as I did through EarthCam, I must ask myself the degree to which I am viewing the physical versus the digital. This is not a new debate. In 1936, focusing on a comparison of photographs and paintings, philosopher Walter Benjamin discussed the benefits and limitations of mechanical reproduction. "Even the most perfect reproduction of a work of art is lacking in one element: its presence in time and space, its unique existence at the place where it happens to be."[3] On one hand, art requires that we stand in its presence to fully appreciate it. We see a piece of art as authentic only when in the presence of the original. However, newer media disrupt the traditional concepts of originality and authenticity. As Benjamin notes, a copy of a photograph (or a DVD of a movie for that matter) is as much the original as the first print ever made. The nature of the way we view mechanical reproduction and authenticity in art translates to contemporary discussions of space. Places share a common characteristic with original art. They have presence in space and time. Do we see space as authentic only if we have visited it, or can an authentic experience of space also be realized in new ways? This question rests at the intersection of digital technology, space, and experience design. Digital reproduction of physical experiences affords us the opportunity to consider the role of proxemics in mediated experiences.

In the days since the live giraffe birth broadcast online, I've asked the would-be zookeeper in me if that was an authentic experience. I've learned that during giraffe births in captivity, zookeepers typically monitor the birth by camera in an effort to stay out of the way to avoid injury to the mother and calf. So, on the day of Kiko's birth, I—and the many others who experienced the birth live via EarthCam—was having the same experience as the zookeepers at the zoo. Yes, they were watching the same event live from the next room while we were watching it all over the world, but our presence in time together in the shared space of the EarthCam live feed made the experience equally authentic for all who were there. We saw it at the same time through the same medium together. The mechanical reproduction of the giraffe birth[4] via EarthCam confirms the importance of not only shared space but also shared time in the discussion of digital proxemics.

Barriers to Entry

Inhabiting spaces is hard. Boundaries that define spaces have ways of repelling would-be intruders, making spaces that we might like to visit inaccessible. And every space and its boundaries are different. When I was trying to conceptualize the boundaries that might keep us from inhabiting a space, the types of barriers to access I considered ultimately fit into three major categories: environmental barriers, systemic barriers, and personal barriers.

Environmental barriers are physical features of a space that make it impenetrable to entry. When my family and I spend a week at the beach, each day we build a sandcastle. My favorite things to design in a sandcastle are its fortifications: the height of the towers with vertical walls, treacherous boulder formations on gradual inclines, guard towers at entry points, and, of course, the basic foundation of any good sandcastle, the moat. My daughter, on the other hand, is not concerned at all about defenses. Her primary concern is access. She consistently discusses roadways and how the kings, princesses, and horses would be able to get up to the castle and live in its towers. She wants to build staircases, bridges, windows, and doorways. During each castle construction project, we go back and forth between access and defense. She builds roads and I position guard towers to protect them. Environmental barriers certainly include the fortifications that might protect a castle such as territorial boundaries created by guards, moats, fences, hedges, or other territorial markers, as well as dangers related to height, terrain, or physically harmful conditions. But environmental barriers might also include proximity and size or scale. A particular space may be too distant to access, or I might be too large or small to access certain kinds of spaces. A space may also be completely hidden from sight, or a memory from a distant time, long since demolished, renovated, or replaced. All of these environmental barriers create physical challenges for those wishing to enter.

Systemic barriers are symbolic features conceptualized for a space that create a culture of impenetrability. These might be realized as rules, regulations, and policies for users of a space; permissions and access codes for entry into a space; or cultural expectations about privacy, disclosure, rights, or responsibilities in a particular space. These barriers might be found in the signage or observed behaviors on site or in the routines, rituals, stories, or myths associated with place. Note that these barriers can come from a number of sources—owners, planners, facilitators, designers, participants—but all are tied to perceptions about how a space should be used and who is allowed to enter. Typically, systemic barriers depend on the assumption of power and authority if the behaviors they govern are to be realized.

Personal barriers are factors present in particular individuals that create obstacles to entry. Personal barriers run the gamut of individuals in society. For one individual, impediments to entry might be the result of body height or size, physical dexterity, disability, or infirmity. For another, entry might be barred because of lack of knowledge about or ignorance of the space in question. Another person might choose not to enter a space based on attitude, beliefs, values, fears, or bias. Note that personal barriers can be real—astronauts had to stand shorter than 5 feet 11 inches to fit into the Mercury space capsule[5]—or perceived. I had a friend in graduate school who refused to shop in a particular big-box retailer because of her fear of birds. To her credit, I had never really looked up to notice the number of birds that lived inside the building among the rafters. When I did, I was surprised at how many were there. Whereas she would consider these birds as a real environmental barrier to entry at this store, I recognize her fear of birds as a perceived personal barrier.

Environmental, systemic and personal barriers can work alone or together to protect the boundaries of any space we might imagine. At a waterpark, the sign at the entrance reads that all swimmers must be 42 inches tall to ride waterslides. This barrier could be a combination of any or all types of barriers. The policy (a systemic barrier) functions alongside the height of the would-be rider (a personal barrier) and is enforced by a lifeguard (an environmental barrier). As you might note, when barriers are merged together like these, a change in one changes all three. A lifeguard who disregards policy lets anyone ride. A person deemed too short could stand on tiptoes to overturn the other two barriers. And, if the policy changes, the other two barriers likely diminish as well.

Digital technology can function alongside existing environmental, systemic, and personal barriers by adding additional protection, encryption, or access points to spaces. These additional protections can provide interesting proxemics discussions about how we move our bodies. For example, many childcare centers have adopted the fingerprint scanner as a vehicle for security during check-in and check-out. From a proxemics perspective, this changes the idea of the queue and the bodily motion of queue participants. But, the technology functions as a new form of physical barrier. Thus, the conversation surrounding the reinforcement of barriers using digital technology primarily concerns the barrier and the way technology is constructed as an additional protection. This type of discussion quickly turns to the technological design of the barrier and not its intersection with spaces. For this reason, the conversation about reinforcing barriers is best left to cryptographers. Readers with a keen interest in this area might turn to Simon Singh's *The Code Book*[6] as an entry point into the vast world of cryptography that goes on behind the scenes of our everyday lives. Instead of focusing on the conversation surrounding additional digital barriers, I am more interested in the removal, disruption, or remaking of barriers that impact our experiences with and in spaces.

The barriers discussed herein shape the ways we move. When we eliminate or alter physical barriers using digital technology, our proxemic interactions are changed. Digital technology can function as an eliminator of environmental, systemic, and personal barriers so that people might gain access to forbidden spaces, remaking inaccessible spaces as accessible ones. By negating barriers or creating wholly digital spaces, which come with new sets of barriers, digital technology can open spaces for a variety of uses and experiences. The remainder of this chapter will first consider the impacts of digital technology on existing physical and hybrid spaces and then the role of digital technology in the creation of new kinds of fabricated virtual spaces we have yet to fully imagine.

Negating Barriers through Digital Access

Although its creators only implied it as such, the first webcam was an experiment in proxemics. Created in 1991 at Cambridge University, the now famous Trojan Room coffee pot camera was designed to make coffee access more egalitarian. Workers in the computing lab realized that those closest to the coffee pot had a better chance of knowing when coffee was ready, so they designed a camera to make the pot visible to all lab members. The camera negated the distance required to see or smell fresh coffee brewing. Writing about the camera, designer Quentin Stafford-Frasier noted, "The Net, once again, helps break down the barriers of distance (even if that distance was only measured in yards), and so streamlines the distribution of a resource so vital to computer science research."[7] The coffee pot's biography explores some of the functionality the camera provided: "The image was only updated about three times a minute, but that was fine because the pot filled rather slowly, and it was only greyscale, which was also fine, because so was the coffee."[8] Stafford-Frasier quipped that the coffee pot's Internet fame didn't make the coffee taste any better. The Trojan Room Coffee Pot camera was deactivated in 2001 and the coffee pot was auctioned on eBay as an item of historical significance, earning the lab £3350 which was used to fund enhanced coffee facilities in a new computing lab.

In the time since their origins in the Trojan Room, webcams continue to shape the ways we move. Webcams and Internet protocol (IP) cameras

connect camera technology as networked devices, able to be accessed remotely and in real time. They allow us to be voyeurs of spaces that we cannot
see physically. Researcher Hille Koskela has termed our present moment in
time as "the cam era."[9] Indeed the camera is the primary vehicle for negating
barriers to spaces. The cameras in our society watch spaces public and private.
As Koskela argues, this pervasive surveillance is incredibly powerful, but the
power of surveillance is likely distributed to multiple watchers. Some cameras are controlled by those with authority to regulate the spaces they watch.
Other cameras are controlled by curious inquirers who seek to observe. And
cameras shared online are likely watched by multiple people with varying motives and interests. Shared cameras like those in the EarthCam network make
surveillance an opportunity for everyone, democratizing access and offering us
all the role of watcher.

One of the early pioneers of the commercialized webcam was Jennifer
Ringley, creator of the JenniCam. Ringley set up her JenniCam in her
Dickinson University dorm room in 1996 and later monetized access to her
live feed. In two years, and following her graduation and move to Washington,
DC, an estimated three to four million people watched her live feed daily,
peering into her most mundane as well as her most private moments, all vol-
untarily made public.[10] Ringley's success with a webcam coincided with an
emphasis in pop culture on satirizing the fame and perils associated with pervasive surveillance, with film releases like *The Truman Show* and *EdTV* and
the origins of the *Big Brother* television franchise.

As the contrast between the Trojan Room coffee pot camera and the
JenniCam reveals, the decision to set up a camera may be rooted in the desire to watch a space or the desire to be watched in space. When we watch
spaces online, we are digital voyeurs seeking gratification from being in places
beyond our own. When we broadcast our spaces online, we become digital
exhibitionists seeking gratification from exposing ourselves to voyeurs we invite into our space. As I write about this contrast, it occurs to me that the role
of the camera for the digital voyeur might be that of negating environmental barriers—removing the physical boundaries to gain access to inaccessible
spaces. Likewise, the role of the camera for the digital exhibitionist might be
that of negating systemic barriers—removing rules, norms, and regulations
that might protect our spaces from would-be watchers. When we engage in
mutual viewership and network directly with others through digital technology, the role of the camera and keyboard is that of negating personal barriers—removing the attitudes, beliefs, and biases to gain access to shared spaces.

Whereas the lines between these three types of barriers are not always clear or distinct, each type of barrier offers us a frame of reference for understanding how we share and move in digital spaces.

Negating environmental barriers with digital technology requires the technology to provide an access point within the barrier. In some cases, this involves placing the camera inside the environmental barrier and allowing it to broadcast out. At the Norfolk Botanical Garden in Virginia, the Eagle Cam has followed a family of bald eagles since 2005. Viewers of the live cam have watched 19 eaglets over that time across several nesting sites in tall trees. The physical barrier of the remote locations of the nest—90 feet off the ground in the tops of three successive loblolly pines—provided the first challenge of camera placement. In 2007, the camera was upgraded to include infrared technology for viewing the nest at night, another environmental barrier. Later in 2011, high-definition cameras were installed. Throughout this time, while watching the live feed of the nest, viewers could participate in a chat with other viewers. This chat erupted in April 2011 when the mother of three eaglets was killed in a collision with a landing airplane at a nearby airport. The live video continued to show the eaglets alone, calling for food. Viewers in the chat room called on the Wildlife Center of Virginia to intervene, which it did by removing, raising, and releasing the three eaglets. In October of that year, the botanical garden dedicated its eagle plaza to honor the memory of the killed mother eagle. During the dedication, onlookers spotted a male bald eagle flying over the plaza. The interactive space created by the viewers in discussion about the physical space of the nest resulted in both a physical memorial for the eagle and a digital archive of her life.[11] The interactive space built upon the fixed feature of the Eagle Cam. As fixed features, static, location-based cameras provide entry points for observing and connecting with spaces from behind environmental barriers.

Sometimes additional technologies amplify the static camera's ability to collect data. These include infrared and high-definition, but also microscopes and telescopes. In March 2000, Katie Couric reminded us that the medical industry has been on the forefront of using technology to access small spaces in our bodies. After her husband died of colon cancer, she underwent a colonoscopy on NBC's *Today* to raise awareness for colorectal cancer prevention.[12] The colon is certainly too small a place to inhabit, but digital technology can magnify its scale so that the space can be visited visually. Conversely, images shared from the Hubble telescope inaugurated a new public curiosity in the vast expanse of outer space. In 2015, the New Horizons satellite broadcasted images of Pluto[13] during its planned flyby, igniting a flurry of conversation about the dwarf planet.

The satellite flyby and the colonoscopy remind me that allowing cameras to move offers a more dynamic intersection with proxemics. Whereas these two examples speak to places we cannot visit, mobile cameras as dynamic features are creating as-yet-unknown challenges for our inhabitation of physical spaces. As I write this book, an ongoing conversation about the future of drone technologies is captivating industries, and delivery service providers in particular. On July 17, 2015, the first government-approved drone delivery occurred in Virginia, carrying medicine from a pharmacy to a clinic in Virginia.[14] Drones rely on physical line of sight or digital cameras for successful operation. While the use of delivery-drones is currently regulated by the US government, the use of drones for private, recreational, and photographic use remains relatively unencumbered.[15] Drones carry cameras, and when those cameras broadcast live images, as they are now capable of doing, new vantage points of spaces can come alive in real time, offering new approaches to proxemics, both as a result of moving cameras and also as a new sort of dance with moving cameras as one partner.

Live cameras connected to a variety of other technologies—infrared for night viewing, drones for aerial vantage points, satellites and microscopes, and many others—negate environmental barriers and create new types of dynamic features that allows us entry to forbidden spaces. But, these cameras also emerge as participants in our experience of the spaces we now inhabit. The choice about what to broadcast, however, remains with the camera's owner or operator. The camera owner is responsible for the choice to negate systemic barriers.

Negating systemic barriers is the simple act of sharing spaces that we control, and inviting others into them. By allowing viewers permission to see private spaces and private events, users offer permission as a vehicle for transparency, fame, attention, infamy, social connection or any number of other motivations. Overriding systemic barriers through live digital video has shaped the daily work of news reporting, educating, church-going, and maintaining family relationships; ushered in new opportunities for medical care, home improvement, and customer service; and catapulted industries surrounding virtual worlds, multi-player video games, and pornography. Systemic barrier negation has also led to some of the Internet's more controversial events, among them the Edward Snowden leaks in 2013, the hack on Sony emails, 4chan user Toaster Steve's live attempted suicide, and countless examples of illicit content shared online through every possible outlet, from Jennifer Ringley's JenniCam to Congressman Anthony Weiner's self-exposure on Twitter. These events and the choice to share them upend the rules of systemic barriers.

In contrast, personal barriers are negated among individuals in the creation of gathering spaces separate from and connected to the physical spaces inhabited by each individual. These sociopetal spaces can be created by webcams but also through a variety of mediums, including text-based interactions (synchronous chats, dedicated forums for asynchronous connection, SMS and text messaging), video-based interactions (Skype, FaceTime, Google Hangout, among a variety of other options), multi-modal interactive broadcasting tools (Meerkat, Periscope, and live webinar technologies among others), and virtual worlds (Second Life, massively multi-player online role playing games [MMORPGs], and other avatar-based encounters). Avatar-based interactions in virtual worlds and MMORPGs are worthy of a brief discussion as they relate to proxemics. Dmitri Williams at the University of Southern California's Annenberg School of Communication has theorized what he calls the mapping principle.[16] His mapping principle is an inquiry of the ways that and the degrees to which nonverbal communication patterns replicate themselves in virtual worlds.[17] When we enter virtual worlds as avatars, we are confined to the programmed digital movements designed into the world we enter. However, the personal barriers users face as a result of their physical condition are overturned, only to be replaced by barriers associated with both the visually salient and available cues in the new virtual world and the user's knowledge of how to operate in the new virtual world.[18]

Like Williams, others have studied the breadth of modalities mentioned above, led in large part by the work of communication researcher Joseph Walther.[19] Dr. Walther's body of research advanced the discussion of computer-mediated communication as a viable site for interpersonal interaction. Among other noteworthy ideas, Walther theorized hyperpersonal communication, the idea that the interaction we experience in online settings might even exceed face-to-face interaction because of strategic advantages afforded by technology. The variety of these types of tools speaks to our collective desire to use digital technology to overcome barriers to communication. As many researchers have noted, these sociopetal digital environments from live chat to virtual worlds can enhance connections for all people in marginalized or stigmatized communities, and these environments can also be a training ground for education in digital literacies. The role of access, education, information, and training related to digital technologies is vital to negating personal barriers and participating successfully with others in the spaces we inhabit.[20]

In terms of proxemics, time is a crucial component in the opportunities to negate barriers through digital technologies. Because proxemics requires inhabiting a space with co-present others in space and time, barriers must be broken in real time and participants must experience the event live to engage in digital proxemics. But, the spaces may very well be physical, digital, or a hybrid of the two. And, the co-present others in each of those spaces all experience proxemics differently.

Earlier in this book, I made the claim interactions in space might be influenced by the space to be sociopetal or sociofugal. In addition, I offered that experiences in spaces might be proxemopetal or proxemofugal. Webcams and virtual worlds stand ready to upend our understanding of these concepts. Webcam technology is simultaneously sociopetal and sociofugal, proxemopetal and proxemofugal, each from a certain point of view. The variety of points of view come from the variety of intents of the user in entering the spaces they inhabit.

Cameras have this funny way of capturing our lives. In one circumstance, they can preserve a memory in time and help it crystallize. In another, they can connect us together in real time, creating new virtual spaces for interaction. This issue of time resurfaces when discussing these barriers and digital proxemics. In these three cases, we're interested in the negating of barriers that connects people to people or people to places. Herein we see that the user's experience, the space, and the technology cannot be unraveled in the context of the proxemics interactions that are occurring. Proxemics in the digitized world must consider space, technology, and experience. The socio- and proxemo-spatial characteristics of the environment and the user offer insights into the nature of this intersection.

In the digital world, a sociopetal experience that is also proxemofugal is perhaps the experience of the digital exhibitionist. The exhibitionist is oriented toward the camera, toward the audience, but is projecting her present environment out through broadcast. Even though the exhibitionist is physically present in space, she is not connecting to physical space. Instead, she is broadcasting her space to be experienced by and connected to others. In contrast, a sociofugal experience that is perhaps also proxemopetal in the digital world is the experience of the digital voyeur. The watcher is not oriented toward social interaction, but only toward entertainment. The voyeur connects with the space created onscreen, despite the fact that such space is not his to occupy. While the voyeur may connect with an on-screen other, such connection only occurs in his mind. While some thinkers may find these two categories a fertile ground for thought and investigation, I am far more interested in their complementary categories, which demonstrate an alignment of people, technology, and spaces in shared experiences.

In the digitally created environment, for the experience of space to be both sociopetal and proxemopetal, multiple users must be fully immersed in the same spatial experience together. In this scenario, two key elements must come together. First, multiple users must be acting together. Users must be connecting. Second, the immersive nature of the space must be shared together. The space they are connecting to is fully digital, either co-created by mutual live stream sharing or manufactured by a third party application. Users might experience this scenario in an avatar-based virtual world, but it might also be realized in programs with live chat functions like Google Hangout, Skype, or FaceTime. Conversely, an experience that is simultaneously proxemofugal and sociofugal is the experience of the immersed in relation to co-present others that are not immersed. The immersed user is connected to the virtual space and to virtual others, but disconnected from the physical location and co-present others. Certainly, co-present others are the first to note this disconnect.

From this point of view, the use of these technologies can be highly sociopetal, orienting people toward each other. These technologies can also be highly proxemopetal, orienting people toward the digital spaces they create together. The role of the camera in negating barriers in these examples above is to show real spaces to people in other spaces, and invite them to participate. However, participation in digitized spaces is not limited to digitized captures of physical locations. New immersive digital and virtual environments spaces invite us to rethink the entire concept of the digitized space.

Remaking Barriers through Virtual Environments

In what is perhaps the most blatant example of disruption to these barriers, virtual environments stand poised to ignite a new method for spatial interaction. A virtual environment can be defined as one characterized by telepresence. Media researcher Jonathan Steuer describes telepresence as the experience of being in an environment through a communication medium. For Steuer, telepresence is a function of both immersion and interactivity with the environment.[21] The feeling of telepresence is a forgetting of the present physical environment due to the vividness of the manufactured virtual environment. Immersion and interactivity are components of the experience that rely on both the user's willingness to participate and the technology's ability to simulate the environment in real time.

Immersion, according to Steuer, requires breadth and depth of experience. In terms of the depth of information, the user must be able to focus on what he chooses in the environment and peruse its details. So, each element in the landscape must be able to be experienced successfully. Breadth of information relies on the interplay with other senses. Current virtual devices privilege visual and aural sensations over the remaining senses, but greater immersion could be achieved through the addition of other senses to the experience. Interactivity, Steuer suggests, is a function of speed, range, and mapping. Speed is akin to a refresh rate, how quickly the computer recognizes user behaviors. Range references the ability of the computer to allow a broad set of options for user behavior. Mapping refers to the ability of the device to accurately respond to user movement. Any upset in the interactive process creates latency, a gap between user action and computer response. For an example of latency, imagine turning your head to look to the left and having to wait for the scenery to catch up. Latency is typically considered an undesirable element requiring correction, as latency can cause disorientation, motion sickness, or loss of balance.

Devices that create virtual environments control and constrain immersion and interactivity to create an experience for the user. The success of the virtual environment relies on the computing power of the device, its graphics card, and its sound and visual displays to provide the intended experience. In this way, environmental and systemic barriers are remade through the hardware and software designed for the experience. Virtual reality (VR) devices rely on head-mounted displays that monopolize the field of vision, projecting a user into another space and augmenting the space with virtual elements. In 2015, technology companies jockeyed to produce and market VR head-mounted displays in the consumer marketplace. Among the contenders, the Samsung Gear VR was the first to release, followed by planned releases of Microsoft's HoloLens, the Oculus Rift (owned by Facebook), and Sony's Project Morpheus. Meanwhile, plans for VR headsets made from old cardboard boxes were released by Google, allowing people to insert a mobile device into the headset to create a VR experience. While Google Cardboard may be a good entry point to the VR experience with existing technology, electronic VR headsets and the platforms they support are poised to create virtual, hybrid, and augmented realities that could reshape education, entertainment, gaming, customer service, and any number of industries that connect people.[22]

As a medium, virtual reality is all about space. VR technologies aim to immerse users in spaces so fully that we experience the virtual space without worrying about the physical space. The navigability and manipulation of the virtual environment are inherent components of interactivity. At the 2015 E3 conference, the mega-conference of video game design, Oculus announced the combination of the Oculus Rift headset and the Oculus Touch, a handheld motion-capture device that works with the Rift.[23] The Touch captures users' hand motions in space and allows them to manipulate the virtual environment and its virtual artifacts through a series of sensors, buttons, and triggers. These types of devices remind the user that they are inside a computer-generated environment, but they allow for more interaction with the virtual environment. While many of the handheld devices, like Touch, are currently clunky and come with a learning curve, their designs and applications will only continue to improve and become more intuitive.

To enhance user ability to manipulate the virtual environment, virtual reality is also inspiring changes in physical space. To move in virtual environments, the physical space actually occupied comes with one particular environmental barrier. It must be open enough to allow free movement. Once the virtual world became immersive, it immediately required a physical space to house the motion of the body. The immersed body moves around the virtual space, unaware of its physical surroundings. Think about immersing yourself in virtual reality in your living room. Your mind would see the virtual world, perhaps a forest where you could search for wildlife. As you move your feet to walk down a path in the woods, you can avoid running into virtual trees, but you would neither see nor be able to avoid running into your sofa, end table, floor lamp, or wall. So the physical location of the VR experience requires the user to remove from his environment anything that would obstruct intended motions in the VR experience. I'm imagining people walking around their living rooms and tripping over their dogs. The digital becomes physical once again. As a result, virtual reality experiences will require relatively open spaces. Virtual environment designers don't see this as an obstacle for the average consumer. Powerful technology has the ability to change the way we see, value, and use our own spaces. After all, as Vive headset designer Jeep Barnett noted,[24] people added garages to their homes to enclose cars, so they might also be willing to rethink their living spaces for another valuable technology, like virtual reality. Barnett suggested that by removing dining tables, selling pool tables, or reconfiguring beds, rooms devoted to large pieces of furniture might be good options for rethinking living spaces to include enough space for virtual reality in our homes. Virtual reality could give a whole new meaning to the current trends of man caves and media rooms.

Other entrepreneurs are developing their own solutions to the need for open space, one of which is the virtual reality theme park. The rise of virtual reality theme parks as sites for creating virtual spaces overtop of blank physical spaces is a fascinating study of the physical-digital intersection. Entrepreneurs across the globe are designing virtual reality theme parks in hopes that people will want to congregate physically to connect together in virtual environments. The first of these, Zero Latency, opened in August 2015 in Melbourne, Australia. Zero Latency uses the Oculus Rift and Sony PlayStation Eye cameras to connect players in the gaming interface. The Zero Latency facility boasts 4,300 square feet of gaming space for up to six players.[25] The Eye cameras, 129 of them, locate players on a grid to map their movements in physical space. As people move in the blank room, their avatars are digitally inserted into the appropriate places in the virtual environment, allowing all players to see the avatars of their co-inhabitants and interact in the virtual space.

In Salt Lake City, Utah, entrepreneur Ken Bretschneider is betting on virtual reality theme parks.[26] When completed, The Void, Bretscheider's gaming space, will be a series of empty 1,200-square-foot pods with foam walls. Each pod will allow 10 players to operate together in a virtual gaming platform layered on top of physical spaces. Physical artifacts in the space will be added to mirror the landscape of the game so that a player might engage in passive haptics in the virtual environment. The virtual designers' term for touch associated with physical artifacts placed in a virtual environment—passive haptics—might include moving around actual walls, grasping actual objects, or feeling actual steam or heat when those items appear in the virtual environment, for example. This type of system requires designers of spaces like The Void to create or customize their own headsets, games, and environments. But that also allows The Void and other spaces like it to provide a unique set of experiences, unavailable anywhere else.

As philosopher Walter Benjamin noted, mechanical reproduction brings into question the concepts of authenticity and originality as contingent upon being there in the presence of the original. As I write about virtual theme parks, I wonder if these types of experiences will become another sort of art to be experienced, complete with a manufactured authenticity of sorts. French philosopher Jean Baudrillard might add these experiences to his list of simulacra. Baudrillard[27] defines simulacra as those copies that have no relation to any original. He cites Disneyland as an example, calling it a simulation of the real world or a fictional account of an imagined reality. In Baudrillard's argument, Disneyland is only a clear example of a fictional reality that we live out on most days. Likewise, virtual environments may be easily recognized as fictional, especially in their early days. But as immersive technologies continue to improve and harness the power of all of our senses, the virtual environment may begin to mirror the real environment in dramatically compelling ways in theme parks as well as in our homes.

For digital proxemics, the questions surrounding immersive virtual environments are many. Will culture or environment serve as the primary author of behaviors in virtual environments? Since proxemic behaviors are indicators of culture, will we bring them with us into virtual environments? If virtual environments assume certain user behaviors that contradict cultural norms, will the virtual environment be strong enough to shift behavior patterns? Will we be willing to act out new proxemic patterns in new and yet-unknown environments? Digital technology shapes the ways we move. With the rise of virtual and hybrid environments, I argue that our responsibility is to understand the technologies that create our experiences. When we understand the nature of our environments—physical or virtual—we become cognizant of being intentional, capable of acting with purpose, and empowered to control our own experiences. Otherwise, we simply let our environments, and their designers, control us.

· 8 ·

DEVELOPING LITERACIES

My family and I were recently window shopping at an art gallery near our home. We enjoy seeing the world through the perspectives of different artists and searching for meaningful original artwork. Among several pieces of art in the window at this particular gallery, the proprietors included a television screen on the storefront that scrolled through images of featured artwork in their collection. As I watched the images move in slowly and then linger in succession, my then three-year-old daughter walked up to the window to get a closer look. Reaching out her arm, she swiped the glass.[1] Much to her dismay, nothing happened. I walked up to encourage her not to touch the television. But then I noticed something curious. On the glass housing the screen, dozens of tiny little fingerprints littered the images. The prints almost all created smears from right to left. Like my daughter, the toddlers in our city and their little hands were trying to interact with a static display by swiping the current image to see the next one. I watched as another child walked up and attempted to interact with the display. He, too, was saddened when it did not respond. Many of the adults observing the scene might have been pleasantly surprised if the television was a touch screen. All of the youngest children on site were surprised that it wasn't.

At the turn of the last century, education thought leader Marc Prensky famously coined the terms "digital native" and "digital immigrant."[2] Prensky's digital natives are those who were born into the digital language of Internet computing. Conversely, digital immigrants are those who are not native speakers of the language of the Internet. Digital immigrants have to learn their way through the digital landscape, but always retain traces of their pre-digital knowledge. The oldest digital natives were born in the early 1980s. But they, too, will become digital immigrants once again, when entering hybrid spaces.

Save the very young who will grow up quicker than we care to realize, we are all digital immigrants in digital and hybrid spaces, learning our way among the physical-digital intersections that we inhabit. These spaces come with modified or new kinds of proxemic interactions that will require us to learn to be a part of a culture unlike our own. We will be living in a foreign world. Many, like Prensky, argue that we are already. What do we do? We live in a time when our bodies are becoming more digital, when our cities are designed to influence our lives, when our refrigerators will order milk for us, when our mobile phones imbue our lives with access to vast information, when we live under a cloud of pervasive surveillance. But for all of us, the hybrid space is here.

This hybridization of spaces should now come as no surprise. We live in a world of fusions. A 1980s canine mutt has become a prized peekapoo or labradoodle. The minivan has given way to the crossover vehicle. The iconic string of Pizza Huts with their unique rooflines and red tile roofs have merged into much less architecturally interesting Pizza Hut Wingstreets.[3] Hybrid American chestnut trees are being bred to withstand bugs and blights so that Appalachian forests may be repopulated with the once common tree. Restaurants boast fusion menus that blend tacos and sushi or southern comfort food with Indian cuisine. The Grammys have added categories for performances that defy genre. Hybrids are all around us. Some, like hybrid cars that run on a combination of electric and gas, have been lauded and incentivized. Others, like GMO hybrids in our food supply, have been met with revolt and outrage. The verdict is still out on most hybrid spaces. So far, our tacit compliance with digital intrusions into spaces implies that we will continue to be propelled onward.

At present, most of our digital intrusions take the form of smart people using sentient objects, or the reverse. Sentient objects[4] are a category of physical object to which some digital connectivity has been added. According to some measures, the number of devices connected to the Internet is already higher than the number of people on Earth. The general consensus among designers of digital tools for hybrid spaces is that approximately 50 billion devices will be connected to the Internet by 2020 and continue to grow exponentially from there.[5] Digitally connected pacemakers, RFID chips in workforce badges, wearable fitness trackers, and smart thermostats are each an example of a sentient object currently in widespread use. In current discussions, these sentient objects are characterized among hundreds if not thousands of ideas under development in the category popularly called the Internet of Things. The Internet of Things, or IoT, is the concept that the physical world and the digital world will blend together in a hybrid state interconnected by technology both human-to-machine and machine-to-machine. In a quick survey of my home, the number of Internet-capable devices typically connected to our wireless Internet service rounds out at a solid 8. This number is likely to grow as more consumer-oriented products compete for our attention.

Recently, the Nest Thermostat has emerged as the consumer frontrunner, leading the charge toward smart homes. Nest Labs, now part of the Google consortium,[6] introduced the thermostat, which operates in conjunction with mobile devices to optimize temperature and energy use. Rumors and speculation—about smart refrigerators that will add milk to our digital grocery lists and smart store window fronts that will note our shopping preferences— continue to fuel discussions about the Internet of Things that is upon us. Researchers at the University of Australia, Canberra, surveyed the Internet of Things marketplace in 2014[7] and categorized IoT devices in the marketplace into five groups: (1) smart wearable technologies, (2) smart home technologies, (3) smart city technologies, (4) smart environment technologies, and (5) smart enterprise technologies. In their study, smart wearables included items like digital fitness trackers and smart watches, among other accessories and clothing aimed at connecting our bodies to technology. Smart home technologies included all manner of Internet-connected appliances, cleaning tools, and cameras. Smart city technologies included fast passes, mass transit tools, and optimized interactive surveillance technologies that mirror the physical city in the digital world. Likewise, smart environment technologies included products that connect us to spaces we inhabit through static devices present in environments and our portable mobile devices. Last, smart enterprise technologies included products that offer business and marketing solutions, use Internet connectedness as a data mine for consumer preferences, or allow consumers to digitally try on or try out products in their own spaces. As these categories of products grow beyond the novelty phase and into common use phases, as some are beginning to do, they are creating hybrid digital-physical spaces that connect to us, the Internet, and any number of potential data collectors. The onus of responsibility remains with us to not only hone our ability to use these hybrid spaces, but also to understand how they work and reflect upon how they shape the ways we move.

Defining Digital and Media Literacy

Contemporary media literacy is rapidly changing. Discussions of media literacy in America have traversed the media landscape from newspapers and advertising, to television, radio and film, to Internet-based media. With new focus turning to literacy in digital realms, several organizations are examining media literacy in its increasingly digital form. These include research centers in universities around the world such as the Center for Civic Media at MIT, the Digital Media and Learning Research Hub at the University of California–Irvine, the James L. Knight School of Communication at Queens University of Charlotte; granting foundations with interests in media, notably the John S. and James L. Knight Foundation and the John D. and Catherine T. MacArthur Foundation among others; and public and private organizations like the American Library Association, MediaSmarts (Canada's Centre for Digital and Media Literacy), CommonSense Media, and the National Telecommunications and Information Administration and its counterparts worldwide.

When I and others in the Knight School of Communication[8] discuss our plans for major initiatives in digital and media literacy, the question I hear the most frequently is, "What exactly is digital and media literacy?" To be fair, the term does require some explanation. In his book, *What Video Games Have to Teach Us about Learning and Literacy*, James Paul Gee defines literacy as the ability to "recognize (the equivalent of 'reading') and produce (the equivalent of 'writing') meanings in a domain."[9] I like his definition. Here's mine: Literacy in a particular medium is the ability to read (or do whatever the equivalent of reading is) and write (or do whatever the equivalent of writing is) in that medium. Literacy is the sum of the processes of assimilation, examination, and generation of information in a media form.

Thus, digital and media literacy involves any number of digital "reading" and "writing" techniques across media forms. These media might include (but are not limited to) words, texts, visual displays, motion graphics, audio, video, and multimodal forms. Literate users of technology will be able to consume and create digital compositions. They will be able to evaluate information on the issues of credibility, reliability, and honor; and make judgments about when and how to apply information to offer solutions. As I shared this definition with a group of community leaders one Friday at our university, one rather accomplished observer remarked, "I note that this definition of literacy renders me completely illiterate." This is the challenge we all face together. Literacy in its many forms is a critical tool for a civil, productive society. The responsibility falls to each of us to seek out and learn to use the tools necessary, and to share those tools with others.[10]

Each of us comes with a unique subset of knowledge, skills, preferences, and abilities to apply to digital technologies. And we are all expanding those capabilities through the experiences we encounter everyday. Our role in an increasingly digital world is to advance our own digital literacies. In *Digital and Media Literacy: A Plan of Action*, professor Renee Hobbs described digital literacy as a "constellation of life skills that are necessary for full participation in our media-saturated, information-rich society."[11] Among these skills emerge five concepts for digital and media literacy: (1) the ability to access and understand information and ideas shared digitally; (2) the ability to analyze online information to ascertain its quality and credibility; (3) the ability to create multimodal content across media forms; (4) the ability to reflect on one's actions in the online environment; and (5) the ability to act in socially responsible ways by participating as a member of a community. Hobbs advocates for schools and education centers to teach these skills to today's learners. *The Journal of Digital and Media Literacy*[12] explores these five concepts as vehicles for community engagement.

In his article for the *EduCause Review*, futurist Howard Rheingold described a set of five literacies needed for life in the 21st century.[13] His list includes (1) attention, the ability to direct and focus personal attention; (2) participation, the ability to inform, persuade, or engage with others; (3) collaboration, the ability to work with others toward a common goal; (4) network awareness, the ability to engage with social, technological, and educational networks; and (5) critical consumption, which he also terms crap detection, the ability to recognize credible information when we see it. According to Rheingold, the possession of the five literacies working together allow a person to effectively participate in our digital culture. Like Hobbs, Rheingold argues that learners need these skills to progress in the rapidly changing information society.

Both Hobbs's and Rheingold's concepts of digital literacies speak to the need for people to participate in the design of their own experiences. When translated to spaces, we are reminded that a typical interaction with space is to accept the conditions of that space and allow the space to shape our behaviors in it. Robert Sommer called this a "fatalistic acceptance" of the environment that "hinders" our interactions.[14] As digital technologies and spaces continue to intertwine, we must consider what digital literacies might look like for interactions with and in hybrid and digital spaces.

We will all be challenged to hone our ability to communicate with and through digital technologies in our spaces in the years to come. Hybrid spaces, digitized places, and digital spaces are neither good nor bad. Technologies of all kinds come with benefits and drawbacks. Enriched uranium can produce energy to power cities or create bombs to destroy cities. Hammers can pound nails that build houses or break windows to loot business. Books can spread truth to promote understanding or disseminate false information to spread lies. Likewise, technologies in our spaces can enhance our experience through better design of optimized environments, democratization of power, and access for marginalized people. Or, technologies in spaces can detract from our experience or be used against us through creation of new power structures, control of access to information, or propagation of dishonesty resulting from hacking, false information, or malicious intent. As digital citizens in smart environments, we have the opportunity to be a part of the environment and choose the actions we will take. My hope is that we will choose a path defined by knowledge, personal ownership of our actions, intentional interactions with and in spaces, and concern for the well-being of others. Digital literacies for spaces will help us to do that. In the spirit of Hobbs's and Rheingold's 5-point plans for digital literacies, I offer five literacies for navigating digital and physical-digital hybrid spaces.

Presence

Being present in a space, being there, is perhaps the most challenging of all of the literacies and the most foundational to our success as humans in a mediated world. Being present means that we are focused on and attentive to the space we inhabit. Digital technology has a keen ability to divert our attention from space to technology, so we must practice the skills of being present. When we need to be present in a physical space, this means shutting down elements of distraction—physically if necessary—so that we can focus on the space and event at hand. This often means powering down a mobile device for the good of a face-to-face interaction. But, it might also mean using headphones to tune out the distractions in the physical world in favor of participation in a digital interaction. Or, it might mean quieting external distractions while competing in a video game. Being present in space hinges on this ability to tune out or turn off distractions to optimize participation in time.

I think about participating in a Twitter chat, for example, and how easy it is to simultaneously turn my attention to my email, a friend wanting to video chat, backchannel conversations on Twitter, or listening to the conversation at the table next to me at Starbucks. So many things—digital and physical—compete concurrently for our attention. AOL Instant Messenger was the connecting technology du jour several decades ago. I remember how upsetting it was it to be chatting synchronously with someone and then, suddenly, even in mid-sentence, she disappeared. Was she choking on her dinner? Did someone kidnap her? Or, in the worst possible scenario, did her dial up Internet disconnect? The wait to find out was horrible. But it could just be that someone walked into the room she was in, or she received a phone call, or she was chatting with someone else at the same time and forgot to click back on our chat. This experience in present day operates more frequently in the converse. We've all had the experience of sitting in a synchronous, face-to-face interaction with a friend when a text message alert pops up on his mobile phone. I'm in mid-sentence. He receives a text message, reads it, writes a response, puts his phone down. By now I've finished my sentence, which prompts him to say, "Sorry about that. What were you saying?" We've probably been the perpetrators of this lack of presence. My point is that this lack of presence can happen in the digital space and the physical space. Our goal should be to select the space we currently occupy in time and be present in it, as fully as possible.

The skills of presence might include selecting when to respond to digital alerts for emails, text messages, or phone calls; making eye contact; developing comfort with video conferencing; employing thoughtful and well-timed physical and digital "away messages"; using digital tools to prioritize incoming messages; developing personal time management strategies; or learning to orient our bodies toward our selected space (and away from alternate spaces). Each of these examples relates to our ability to be present in space, but perhaps more important, present in time.

Judging from my observations of a wide swath of our society, myself included, I believe that the skills of being present require practice. Some of these skills might require that we each adopt personal rules of engagement.

For example, I might have a self-imposed rule that says, when talking to a co-present other, my mobile phone is dimmed and not in my hand. Another might be, if I am in a video chat, I close my email window. Unfortunately for us, adhering to these types of rules does not come naturally. We are built for multi-tasking, and in many cases can do it well. However, our ability to focus and target our attention in the spaces we encounter is a skill we should claim as important and hone over time.

The concept of being present in space also raises the issues of entrance requirements. We cannot be present in a space to which we do not have access. In chapter 6, I discussed geo-fencing and its relationship to place-based, digital access to information. I was recently driving down a busy road when an alert sounded on my phone. When I later reviewed the alert after safely parking my car, I saw that Groupon, a web-only coupon application, alerted me that a deal was being offered on its site at Krispy Kreme Doughnuts, one of the stores I passed on the road. The information presented was place-based and timely, if not also relevant. But, it occurred to me that everyone else on the road that day may or may not have received that same information based on their access to technology, their choice of applications, or their browsing history. Just because the information was tied to that place does not mean that everyone could see it. That discrepancy may be fine, and even preferred, for online coupons, but for documents related to area governance, land disputes, historical archives, or public records, such place-based access may indeed be counterproductive. These are only a few of the underlying digital literacy concerns raised by place-based, digital access to information.

Analysis

The role of space in our lives requires constant interpretation. As with all product-based interpretations, our minds process interpretation on three levels: visceral, behavioral, and reflective.[15] The visceral level is the gut reaction. At a most basic level of survival, our bodies are hard-wired for fight or flight responses when our personal space is threatened—to know when to stay in a space and when to run away. This visceral response shapes the relationships that we have with space, requiring us to constantly sense and interpret the space and the motions of others within it. The visceral response also determines our reaction to a space upon entry—whether we love it, hate it, or are ambivalent toward it. Our visceral response to digital technologies in physical spaces and wholly digital spaces may depend on our comfort with the present technology. And we are divided in how to respond. Early adopters express joy at the sight of a new digital device. Luddites reject the digital completely in disgust or concern. The people in between might register mild concern or apprehension, or they might enjoy the novelty of the experience. These reactions can be helpful assessments of one's self.

When we are not in danger, as is the case most of the time, we can peruse, amble, and orient ourselves in space. While we are in the space, interacting with the space, we interpret the space on a behavioral level. At this level, we seek to understand the culture of a space, the behaviors that are required or prohibited, the ways we might move in the space, or the purpose of the space in our experience. As we use the space, we make judgments about how easy it is to use. This ease of use might be in terms of spatial features, sociospatial design, navigation and wayfinding, or intuitiveness of digital interactions. In digitized spaces, the behavioral interpretation of our communicative messages can shift when digital technologies intersect with physical spaces. For example, researchers Poeschel and Doring[16] note that text-based messages, like emails and SMS messages, can contain information about geographic location. When we receive messages, they argue, the geographical distance and type of location represented can both alter the ways that the message is interpreted. In terms of geographical distance, Poeschel and Doring suggest that a message sent from a very close position to the receiver might be interpreted differently than a message sent from very distant position. Consider the simple example of a text from husband to wife: "I miss you." Geographic location adds an interpretive level to the information in the text. If the text is travelling 300 miles from person to person, the wife might read the message as a wistful and kind anticipation of their reunion. However, the message might offer a different interpretation if this same text, "I miss you," was sent between spouses seated on the same couch. It might be read as a plea for attention, a rude remark, or a signal that we are paying attention to our phones rather than each other. In the several years since Poeschel and Doring suggested this relationship between email and proxemics, geo-location tags have been widely added to social media across platforms. Facebook and Twitter, among many social networking sites, have added location stamps to posts. Most mobile devices add information about geographic location to photo and video files. Search engines also use geographic location as a filter when returning results. These locations become, quite literally, digitized. Our interpretations of spaces hinge on our awareness of how messages and people move through the space in the process of our physical and digital visits.

Once we leave a space, we can start to engage the reflective level of interpretation. This level exists in memory as we think of the space fondly, recommending a visit to our friends, or negatively, fuming over the poor ambience or the lack of ability to read tone in text-based messages. The digitization of spaces requires that we take time for reflection so that we can learn how to interpret the spaces and the messages within. As we reflect on spaces, some become hallowed places of importance in our lives. Others become sites of archived happenings. Still others fade from memory altogether. As we've previously discussed,[17] our archives of memory are shaped and colored not only by our retelling of the stories but also by the way we capture and share these experiences digitally.

The multifaceted nature of our spaces as physical, digital, digitized, blended, hybrid, and virtual offers new avenues for interpretation and new types of cultural practices observable in these spaces. In his book *Mobile Interface Theory*, Jason Farman makes the case that when we use locative mobile technologies, we connect to physical space in a digitized, hybrid form. Arguing that the duality of physical and virtual is perhaps no longer relevant for space, Farman says, "Locating one's self simultaneously in digital space and in material space has become an everyday action for many people."[18] This simultaneous interaction changes our cultural definitions of space and the role of the objects within spaces. As Adriana de Sousa e Silva articulates, the feeling of being in two places at once is enhanced by mobile technologies. Mobile phones provide "the possibility of moving through space while interacting with others who are both remote and in the same contiguous space via one's relative location to other users."[19] In this case, the boundaries between the physical and digital spaces blur for the user, creating a hybrid environment.

Throughout this book, I've discussed many ways to interpret digital proxemics and suggested the inherent value of personal awareness of self in space, commenting on the change in space resulting from technological advancement in digital and Internet-based tools. I won't continue to paraphrase the many examples here, but I will say that learning about my personal use of space—how I move in space and how spaces move me—fundamentally alters my ability to interpret spaces and to make choices about how to engage in and with the spaces I inhabit.

Intentionality

As a scholar who researches and studies space, my typical argument about space is that we should use spaces intentionally. I want to free all of us from the commonly held assumption that our spaces are inflexible, static, and constrained. Spaces are designed and we can select our behaviors inside them. During a visit to London's Buckingham Palace, my friend Adrienne posted about her experience on Facebook, writing, "Ben and I danced for about 20 seconds in the Buckingham Palace ballroom today. Because if we didn't I would have regretted it forever. #defythequeue."[20] Her sentiment reflects the struggle we all have in inhabiting spaces. We have to decide how to act. On one hand, we want to respect the boundaries of the space we inhabit. On the other, we yearn to experience spaces for what they are. A ballroom is made for dancing. And yet, her hashtag, #defythequeue, notes that dancing was not the expected behavior for visitors moving in line. For the record, I would join the dance.

Intentionally selecting our behaviors in spaces allows us to transition from users of a space to actors in space. I am not suggesting that we take on the mantle of architectural possibilism and begin a crusade to remake spaces as whatever we want them to be. Rather, I would like for us to make active choices about the behaviors we adopt in spaces, consider the possibilities, and select the behavior of best fit. Spaces are designed for particular experiences. Sometimes, as in the examples of the US Holocaust Memorial Museum and Walt Disney World's Scuttle's Scavenger Hunt queue, relinquishing our choices and succumbing to a space in an architecturally deterministic way allows us to engage in meaningful experiences. In other times, we may need to act in more possibilistic ways to explore the many options available to us in spaces and choose a behavior of best fit, like dancing in the Buckingham Palace ballroom or disrupting our GPS device's directions to navigate our car around a traffic jam.

Advocating for a critical pedagogy of place, David Gruenewald[21] articulates the concept that we act in places according to cultural practices that may or may not be worth repeating. By learning about and educating others about the places we inhabit, we can claim places through the work of reshaping behaviors and rules that govern the cultural, socioeconomic, and political structures in our places. To develop the literacy of intentionality in spaces, we must learn how to examine and read the cultural practices of the places we encounter and determine which behavioral patterns should be conserved and which patterns need to be transformed.[22]

The act of intentionality is bound up in the ability to direct our minds and bodies toward the desired experience of space and place. As a culture, we need to be more adept at purposefully orienting ourselves in space to enhance our sociospatial experiences. If we want to establish sociopetal presence in an on-line chat, we need to consciously select a sociofugal position in physical space. If we plan to engage with a co-present other in sociopetal physical space, we should be able to orient our bodies in a sociofugal pattern with others in digital space. If we would like to connect to the physical space we inhabit, positioning our bodies for a proxemopetal experience will help us. If we would like to engage in a digital environment, then we need to develop strategies for engaging physical space in a proxemofugal way. The bottom line is that we have numerous choices in the way we position our bodies in space and in relation to others in space. The digitization and hybridization of spaces does not change that fact. Digitization and hybridization of spaces gives us more options for what, who, and when to engage in space. The options are greater in number, but the choice is still ours. And these choices can and should be intentional and purposeful.

Design

The experiences that we have in spaces depend on the features in spaces and our perceptions of them as fixed, semi-fixed, or dynamic. Sometimes, in the case of fixed features, we are forced to comply with the space as designed. However, in more cases than we realize, features that appear fixed are semi-fixed or dynamic, inviting us into the design process. Even though we are unlikely to walk into a space and reconfigure it for our immediate needs, we can. In many cases, we have the power and responsibility to design spaces for our own use.

We will all be designing hybrid spaces incorporating physical and digital components in our homes if not also in our workplaces, third places, markets, cities, and public gathering sites. We will make decisions about the number and types of digital technologies we bring into our homes and offices and those that we wear on our bodies. And the options will be many. In the *New York Times*, contributing columnist Allison Arieff lamented "the Internet of way too many things,"[23] citing products like Leeo[24] (a night light that will call your smartphone if your smoke detector is sounding), Whistle[25] (a wearable fitness monitor for your canine companion), and Mimo[26] (a sensor-laden one-sie sleep monitor that alerts parents about baby's sleep patterns and can connect with other devices in the home to turn the lights on, brew coffee, or play music when baby wakes up). While some users may find these products beneficial—Whistle also has a GPS tracker to help you find Rover if he gets lost—Arieff rightly notes that consumers should not privilege novelty over actual value. She comments that each Internet-capable device we add to our bodies or homes should meet four criteria: (1) it should integrate multiple functions, (2) it should be usable, (3) it should be made sustainably, and (4) it should privilege privacy and security with clear disclosures about what types of data it collects and to whom that data is delivered. These four criteria are issues of product design, but they are also issues of the design of our spaces as we select the digital tools and choose to embed them in our homes.

Designing interactions with technology in our homes raises many questions about how technology functions and how it connects us to the digital world. These questions are amplified when we do not control the design process. As our interactions with digital technologies become increasingly frequent through smart enterprise technologies or in public places and urban centers, we may find ourselves constrained. Smart cities and smart enterprise technologies capture and mine data from us, and our roles seem limited. We can add data to the growing body of data, but how might we contribute to the design process? In *Places Journal*, New School media professor Shannon Mattern calls on citizens to engage in design: "We're empowered to report failed trash pick-ups or rank our favorite hospitals, but *not* entitled to know what happens with our personal data... nor do we know how *our* intelligence translates into urban 'sentience,' and what is gained or lost in the conversion."[27] She raises questions concerning how we interact with the components of the city, where location and access points exist for the interface, how the interface orients users in the city, and who the interface includes and excludes. Her questions again remind us of the power of the designer in the designed artifact. In an ideal world, we want our cities to be designed for its people by its people. The digital infrastructure that undergirds the smart city should be no exception. We are all called to be designers if we want our cities to continue to be places for people.

In the fully digital spaces of the Internet, the components of the literacy of design are bound up in the ability to produce artifacts in digital media forms and make judgments about how and when to use digital tools. We exist in a technologically mediated world that advances those with existing privilege, and limits those without. In the area of design, this relates to the ability to code digital information, produce media objects, communicate through digital media, and connect physical objects to digital networks. How might people develop these design skills? Coursework? Experience? Trial and error? Media researchers Nancy Baym and danah boyd offer, "Some develop a sensibility through experience; others find themselves struggling to make sense of and manage their participation in networked publics; some misunderstand the consequences of their actions and make mistakes without realizing it."[28] The truth is that we've all made mistakes in media creation. We post something to a social network that harms a relationship. We fire off a nasty email that has repercussions for our continued employment. We mistakenly write about private information in a public forum. We author an HTML code incorrectly causing a webpage to fail. An image depicting illicit behavior is unknowingly uploaded to a sharing site by a third party. Despite these mistakes, we are all trying to learn the literacies of writing in media forms.

The impetus for us all to advance our knowledge in the work of design is that our reliance upon digital technology is increasing while our understanding of how technology shapes us is dwindling. It does not matter what skill level of media production I possess today. What matters is that my skills in one day, one month, one year surpass my skills today. We must all be learning continually, creating continually, developing our skills continually.

Whatever our relative skill, our goal for digital proxemics should be to become part of the design process in our spaces. As the Internet of Things becomes more prevalent, our ability to develop the digital literacy of designing interactions will impact the spaces we can inhabit and the ways we will use them. We will bear the responsibility of adding digital components to our physical spaces. And our spaces will demand it of us sooner than we think.

Across all of these intersections between digital and physical, we become designers when we enter the creative process where spaces become digitized. For smart wearables and smart home products, this means understanding where our data goes when collected, who accesses it, and for what purpose. Consumers have a right to this information, and can help businesses design solutions that advance user-centered practice, personal privacy, and ethical action. For smart environments, smart cities, and smart enterprise, we must advocate for a collaborative design process that matches technology developers, business leaders, and technology administrators (who currently hold much of the design responsibility and power) with artists, thinkers, urban dwellers, politicians, and citizens of every kind. Only through a collaborative design process will the realized designs of physical-digital intersections be shared, democratized, meaningful, and accessible for all of us.

Contribution

We must also be honorable people using spaces and digital technologies well. Using them well means that we are contributing our work for the common good. We are not people simply participating in neat digital things, but we are using our strengths to add value to the spaces and people we encounter.

Our goal in promoting the public good must be to use the intersections of digital technology and spaces to add value to lived experience. In the early days of the Internet, many people argued that the digital revolution would democratize information. And they were correct. Information is available in larger volumes and accessible anywhere. However, some media theorists continue to speculate that the digital revolution would also usher in a new set of power players in the political landscape. These designers, writers of computer code, and owners of technology companies are not democratizing forces but rather new players in today's version of socio-political power. As cultural theorist and former software designer David Golumbia[29] notes, computers help citizens, but they also serve powerful interests. Navigating that balance and advocating is the work of the literacy of contribution. It builds on the other four literacies in the service of public good.

Digital technology and spaces can merge in compelling ways to promote the well-being of people for the public good. Consider the "Sense Your City" project.[30] In cities around the world, from Bangalore to Boston, Data Canvas sensors measure sound, light, and air quality as well as collect data on relative temperature, humidity, dust, and pollution. The data produced by these sensors are translated into visualizations that can be used for longitudinal research, personal health information, city archival data, and digital storytelling. Sponsored in part by the Gray Area Foundation for the Arts, a non-profit that supports social good through technology and art, this project creates publicly accessible data that can be harnessed for a variety of purposes, interpretations, and actions. The public good here is not only the collection and visualization of informative data for the general public, but also the openness of the data to all.

This openness of data in cities and public places can also be mirrored in our own lives. Often labeled oversharing, the saturation of our social media sites with self-related information can make us mini-celebrities. As our private lives become more public, navigating our online disclosures will continue to grow as a collective issue. Popular science author Steven Johnson suggests that oversharing might even become a civic good.[31] We have a right to privacy, but selectively opening our lives to readers might become opportunities to inspire others for the better. One of the more researched aspects of this relates to healthcare and education. Health-related blogs, particularly in the case of chronic or life-threatening conditions, have emerged as sources of inspiration. Hope College communication professor Isolde Anderson studied the use of Caringbridge.org sites among patients and authors. She uncovered that psychological support and encouragement were two of the main purposes they chose to share health information publicly. Like Anderson, many researchers have pointed to the role of public writing as a therapeutic process for patients. This is certainly a personal good, but it has the potential to be a public good as well, as people with life-threatening and chronic conditions seek out other patients online to learn from their experiences. In 2008, the California HealthCare Foundation[32] reported on studies indicating that the top three reasons that people use the Internet to connect with others about health issues are (1) to hear consumer reactions about particular drugs or treatments, (2) to research the experiences of others for social comparison, and (3) to learn skills or gain education about a particular condition. In a 2010 Pew Center study of health care and social media,[33] 57% of patients living with chronic disease had connected with user-generated content on social media. The report notes that only 1 in 5 patients suffering from chronic conditions posted health information publicly. This disparity suggests that people are more likely to seek out information than share it, and that those who share their information may be providing a service to others in their similar situations. Audience research on this level is hard to conduct, but the Pew report notes that people commented preferring informational blogs and websites to chat rooms or similar tools. While sharing or oversharing in public online spaces may provide social support for an author in healthcare or other fields, it may also provide information to readers who are less inclined to share. Doctors and community leaders are understandably mixed on whether oversharing is a public good as misinformation can be shared as quickly as correct information. However, oversharing has the potential to connect people in new types of spaces and in ways that serve the good of our society.

In addition to data sharing, contribution to the common good is often the work of activism and emerges through scenes of protest in places. The role of space is foundational to activism as places both house unrest and then stand as monuments to the events that occurred there. Inhabitation and occupation of public places are markers of cultural debates and civil protest. By invitation of the Supreme Council of Culture in Egypt, University of Chicago English professor W. J. T. Mitchell reflected on the role of occupation of spaces in recent political uprisings around the world. In his response,[34] Mitchell examines the intersection of culture, space, and revolution from the Occupy Movement in New York's Zuccotti Park and the National Mall in Washington, DC to the Arab Spring uprisings in Cairo's Tahrir Square and in the tent city on Rothschild Avenue in Tel Aviv—all of which were notably spurned onward through sociofugal applications of social and participatory media. In some of these places, statues stand as iconic monuments and symbols of the right to protest. In others, monuments were erected temporarily in the act of protest. Mitchell argues that the emptiness of open spaces can also be monuments of cultural change. He notes that the famous Champs Élysées and other wide boulevards in Paris were strategically constructed to suppress the types of uprisings characterized by road barricades during the French Revolution. Likewise, an empty Tahrir Square, Tiananmen Square, Red Square, or National Mall might signify a revolution ended in victory, suppressed in defeat, or neglected. People filling a public space might be gathered for a national celebration or to protest against the work of the state.

Similar gatherings of people can also occur in digital spaces to promote public good. For example, Change.org, an online petition site, has over 100 million users across almost 200 countries worldwide who use the site to create and support petitions for causes.[35] These petitions are digital gathering spaces which can create the opportunity for social change. Change.org gained traction in 2011 when a petition from Trayvon Martin's parents recorded a whopping 2.2 million signatures, prompting authorities to respond, gaining national media attention, and eliciting a response from US President Barack Obama. Ben Rattray, the site's founder, created the site not to champion particular causes, but to create a place that allows people to champion their own causes together. Change.org does not approve or assign value to petitions— even though staffers police the site for hate speech and calls for violence. Instead, users of the site support causes they deem worthy by adding their signatures to growing lists of supporters of change. Some might think it important to note that Change.org is a certified B corporation which can both make a profit and advocate for social change like a non-profit. Change.org's revenue comes from premium features and from partnerships with companies and non-profits seeking to target particular groups of users.[36] I'm not suggesting that Change.org acts dishonestly. In large part, all current signs point to the company's desire to be mission-driven toward social change. However, digital technologies have designers and are sponsored by individuals and companies with particular interests. When we use specific technologies to act, we should at least know who is behind both the cause we plan to support and the technology that facilitates that support. The same is true for joining a physical gathering in a public place.

Iterations

As we work to develop literacies for digital-physical intersections, we will all need to become adept actors in the culture of the spaces we inhabit. Spaces are important. They carry cultural weight that illustrates significance, power, and value in tangible and intangible ways. For this reason, we often choose to protect our spaces, adding barriers, guards, cameras, or locks. In American culture, we have placed economic and prestige value on the size of our spaces. Bigger offices, larger homes, taller buildings, and wider expanses of land indicate positions of power, wealth, and authority. One look at an ancient Egyptian pyramid indicates that this prestige value is neither new nor American. Throughout human history, space reflects culture. One might propose, then, that the preservation of space is an attempt to preserve culture. Likewise, a transformation of space is an attempt to transform culture.

Debates over how best to conserve and iterate physical spaces and places are not new.[37] Spaces and places change and deteriorate. New versions of them are created and built over the old. Places are renovated and reinvigorated for new purposes and new times. When I drive past the site of the historic school-house where over the course of a century my grandfather and his siblings, my mother and her siblings, and I and my siblings were all elementary school students, I now see a shopping center with a few vacant storefronts. The school still exists. Its new building is located on some acreage down the road, and the cupola that stood atop the old building is on a pedestal on the front lawn. But it exists in a new place and time. While memories of the former building persist, it is gone.

When the places and spaces in our lives are concerned, myriad factors dictate the conservation or transformation of spaces: ownership, public agreements, deeds, private decisions, finances, water and land use rights, city planning, property values, accidents, weather patterns, acts of God, and the list goes on. This list for physical places is often part of the public record. However, for digital spaces, the items on this list may very well be unknown. The hybridization of the physical and digital space forces us to reconsider how we move in spaces, but it may also force us to reconsider the ways that spaces and places both physical and digital are owned, bought, sold, traded, and zoned. When we are in a space, the way that space has iterated over time and the way it will in the future should be of primary concern to us.

I think about this issue of iteration when I visit one of South Carolina's beautiful barrier islands. These sandy beaches built by sediment, ocean, and weather patterns naturally move and shift. They change location, offering wide expanses of sand in some years and decades of eroding in others. I think about our choices about how to conserve beaches and the role the Carolina coast has played in my life. What I have shared there. What the place has shared with me. How I connected with others there. What I have left there. What I have taken from there as mementos, tangible and intangible. What plans I have for my children to experience there. As the coastline moves and changes over my lifetime, what decisions have shaped it for the better and for the worse.

The paragraph above may make me sound like an environmentalist. And I probably am. I do study space and place, after all. But I wonder what would happen if we started to guard and shape our interactions in digital spaces in much the same way. As digital spaces iterate, how do those changes affect me? How do they affect you? How do they change our culture? Digital spaces can emerge, disappear, or be revised without our attention or knowledge. The Orwellian[38] age of mechanical alteration is available to us on the Internet as both consumers and creators. History can be rewritten from our mobile devices. Our presence in digital spaces can be deleted, erased, amplified, and publicized. As the digital connectedness of space changes over my lifetime, what decisions will shape it for the better and for the worse.

The spaces in our lives, digital and physical, shape the ways we move. We move them. We are moved by them. We orient ourselves according to their cues and our own culture.

The question for us is whether we take the time to notice. And when we do, how will we act?

· 9 ·

RESEARCHING DIGITAL PROXEMICS

From its origins, proxemics has been a tricky subject to research. Even by admission of its originator, Edward T. Hall, human use of space is hidden, nuanced, and tantalizingly complex. In his article, "Proxemics," written for the journal *Current Anthropology* in 1968,[1] Hall outlines the various constructs and methods for researching proxemics that he deemed valuable at the time. His constructs included features of space, sociospatial and sociofugal designs, relationships between proxemics and language, and interpersonal distance-setting mechanisms. His selected methods included observations, experimental abstract situations, structured interviews, analysis of the lexicon, interpretation of art, and analysis of literature. Even more interesting than his article was the response it generated. Paul Bohannan called Hall's research "vital" to human research.[2] Ray Birdwhistell, the designer of the nonverbal concept of kinesics, credited Hall with the "awakening of an interest in the social perception of space which has been semi-dormant,"[3] but later describes some of Hall's ethnographic techniques as "little more than interesting esoterica for the student of communication."[4] Bernhard Bock wrote that Hall's research "calls attention to a number of problems of great importance for the sciences of man,"[5] while A. Richard Diebold, Jr. suggested that the context-specific nature of proxemics interactions adds to the challenge of research.[6] Others responded with a variety of accolades and criticisms (and often both as is the typical case for most academic audiences). J. E. McClellan sums the sentiments behind the critique, writing, "Proxemics has proved an enormously fruitful field of research despite the cloudiness of its defining principle."[7] Hall's work in proxemics was not devalued by this critique. In all likelihood, the importance of his concepts to the study of human experience was bolstered by the attention that it received. As Hall wrote in his reply to the commenters, his exploration of proxemics must have "struck a nerve. It is this sort of event that interests me,"[8] the sort of event that starts a conversation and garners attention for an as-yet-unseen part of our lived experience.

In his article, Hall's list of approaches to research in proxemics was regarded at the time as both inspiring and incomplete.[9] I believe that such is the nature of proxemics. For me, this "cloudiness" makes proxemics fascinating, but it also constrains our collective ability to research proxemics. The study of proxemics, therefore, requires research methods that examine content as well as context, methods that privilege observation of how people actually behave over their own interpretations of the events, methods that reveal patterns occurring across multiple groups of individuals, and methods that allow us to conceptualize movement in multiple, varied, and uniquely compelling spaces. Hall's article led to the later publication of a *Handbook for Proxemic Research*.[10] Hall's *Handbook* chronicles research through the related steps of one intensive research study to offer one possible foundation for proxemics research. Unfortunately, the handbook, now some 40 years old, is unavailable digitally and exists to be uncovered only in university libraries that still include it in their collection.[11] In September 2015, the OCLC WorldCat library cooperative[12] showed only 213 copies of Hall's *Handbook* scattered across the world among the over 2 billion holdings recorded in its system. It is no longer a widespread tool for proxemics research.

Moving proxemics from the realm of physical space to that of hybrid and digital environments does not change the requirements for research. However, digital tools stand poised with the ability to provide researchers with objective data points that can characterize behaviors of individuals, people in groups, and populations in aggregate. Digital tools that create a multimodal archive full of data can then mine that data for patterns and visualize that data as information. And the quality and sophistication of digital devices continue to enhance our ability to observe and capture data connecting users and space.

Borrowing the organizational concept employed in Hall's article, this chapter outlines the researchable constructs in digital proxemics and methods that allow us to examine these constructs in the environments we inhabit. The goal of this work is not to create an exhaustive list of possible research opportunities. Neither is the goal to detail the methodological protocols for conducting these specific types of research. Instead, the goal is to inspire research along multiple methods that can illuminate patterns in digital proxemics and explore the ways we move. With this goal in mind, I offer here a synopsis of the constructs surrounding digital proxemics and introductions to several research techniques that can add value to the conversation surrounding digital proxemics.

Constructs and Concepts

The constructs and concepts included below emerged from a foundation in Edward T. Hall's writing about human use of space and have been reconstituted herein in relation to digital proxemics. The concepts include movement of the body, features and territories, interpersonal distance, sociospatial design, and cultural constructs of space.

Movement of the Body

The relationships of bodies in space—thermal space, visual space, olfactic space, kinesthetic space—were of primary importance to Hall. They laid the foundation for his concept of proxemics. The ways our bodies move in space shape the ways that we interact with others. And, the technology that we add to our bodies has the potential to change both the nonverbal cues we emit as well as our ability to sense our position in space. When technology changes the ways we move, our motions must be able to be reinterpreted in light of the technology applied to them. The movement of the technologized body and its impact on proxemics could be measured by mixed quantitative and qualitative analyses that might keep records longitudinally over time, examine potential amplifications of sensation and nonverbal delivery, or explore comparisons between the technologized body and the non-technologized[13] body.

Features and Territories

The physical and digital components of a space force us to reexamine the territorial features of space suggested by Edward Hall. In his definition, features were fixed, semi-fixed, or dynamic. With digital information overlaying phys-ical features, even the most seemingly fixed features might seem highly dy-namic. As a result, I propose that we add another layer of classification to the concept of features when we consider digital feature space. Whereas physical features might be characterized as fixed, semi-fixed, and dynamic, digital fea-tures might also be characterized as static, interactive, or immersive as a re-sult of their responsive capability, in addition to being characterized as fixed, semi-fixed, and dynamic as a result of the space they occupy.

First, a static digital feature would be a computer-generated component of the environment that does not change. It is perhaps projected on the space or connected to the space. Examples might include a digital audio file appended to a work of art that describes its history, a light display cast on a wall, a geo-locative tag, or a webpage that contains user-generated content about a space. A user might still have to use particular technology to access the fea-ture, but the feature exists as created. The yellow down lines that appear on the football field on television are dynamic features in that they overlay mov-ing boundaries to match the play of the game, but they are also static, lacking user control. Second, an interactive digital feature might be a computer-gen-erated component of the environment that changes or modifies aspects of the environment based upon user entry, exit, or action, or upon user control of the device. The idea of the interactive digital feature is that the user can control and manipulate the feature in the environment. Keep in mind that these are digitally constructed features, not physical features. A physical feature, like a camera or sensor, may be a semi-fixed vehicle for the creation of a digital fea-ture, but it is wholly physical. A visual light display like the one on the floor of the Vanderbilt library is a fixed feature in that the territory it occupies in space is immovable, but it is also interactive, responding to the movements of the user. Third, an immersive digital feature would be a computer-generated component of the environment that reimagines the physical reality of the space. It overlays new boundaries on physical space that exist only in digital space, reinscribing territory completely. However, as noted in the discussion of virtual spaces, current constructions of immersive features dissociate users from the physical location mentally, but the physical features of the space may still win out when the physical and virtual exist in opposition.

These features and the territories they create help us to define the types of spaces we encounter and inhabit. In this book, I've discussed spaces as physical environments, digital environments, digital-physical hybrid environments, and virtual environments. Each of these warrants its own study in relation to the features and territories present in each environment. The classification of features offers us a window into the complex articulation of spaces as conceived geographically, architecturally, and digitally on both micro and macro levels. These might be examined through observations, applied research in environmental design, or usability testing.

Interpersonal Distance

Although Hall suggested that interpersonal distancing was culturally defined, most cultures appear to have proxemic constructs, even if those constructs are different. In his definition, intimate space, personal space, social space, and public space had boundaries that could be articulated by our sensing abilities in space, bounded by cultural definitions. As we have seen, culture in digitized spaces may create a new iteration of cultural patterns or a new cultural system altogether. In addition, digital technologies can allow us to enter new types of environments with unwritten and unimagined cultural rules. At the very least, changes, deviations, and reformations of interpersonal distances can be markers of our changing culture. Interpersonal distance continues to be aptly measured through observations of human interaction.

Sociospatial Design

The positioning of features in space creates opportunities for humans to orient toward each other in a sociopetal orientation or away from one another in a sociofugal orientation. I have added to this concept the idea that digital technologies use spatial and digital features to orient us toward our physical location in a proxemopetal orientation or away from our physical location in a proxemofugal orientation. These two classifications together can help us examine digitized spaces along the dialetics of connection to others and connection to location. Digitized spaces mix these orientations in multiple ways by dissociating geographic location and interpersonal connection. The connection that people feel to each other might be well studied through observations and mixed quantitative and qualitative methods exploring perceptions of nonverbal immediacy, social presence, or interpersonal liking, among others, whereas the connection to location might be studied through perceptions of affinity, presence, and aesthetics, among others.

Cultural Constructs of Space

The impacts of digital technologies on space invite us to reexamine a variety of space-based cultural constructs. Among these are privacy, archives, and surveillance. Privacy, formerly determined in space by the fixed features of one's home and property, has now been reshaped through invasions of digital sharing tools into private spaces. Archives, the spatial storehouses bounded by walls and physical materiality, have now been pushed up to the hazy, semi-protected digital world of the cloud, and pushed back down to phones and devices. The archives that we used to have to travel to can now travel with us wherever we go. Surveillance, the watchman bearing power over the people, surely still exists and is becoming increasingly digital. However, at least in some cases, surveillance appears complemented by the concept of social surveillance, in which people may simultaneously exist as watcher and watched, vacillate between the two roles across time, or select the role most appropriate for their circumstance. The impacts of digital technologies on these constructs invite new examination and longitudinal study. These cultural constructs related to space, among others, might be studied through examinations of cultural values in aggregate, group, and individual settings, critiques of the rhetorical situations surrounding these concepts, and explorations of resistance in its many forms.

Research Methods and Techniques

This list of research methods and techniques can be well-suited for studies of digital proxemics. My aim is to briefly introduce multiple methods with the goal of overviewing a broad set of possibilities for research. To that end, my descriptions are necessarily but purposefully incomplete. I do not intend this to be a research methods textbook, nor could I possibly address the nuance of each method in these few pages. Instead, this list is intended to inspire creativity in the research process and invite a new cadre of researchers to join in the investigation of digital proxemics. Included in the methods below are observation, experimental and quasi-experimental designs, cultural criticism, a brief word about big data as an avenue of interest, and applied research assessments.

Observation

Observation remains a stalwart among research techniques concerning space and human use of space. Hall suggested that the hidden aspects of proxemics could be revealed through observation of the movements of people in relation to each other and the features of physical space. Observation is not simply watching, but includes the exploration of observed data for patterns, exemplars, and outliers. Observational techniques have often been completed with the researcher acting as silent observer or as participant observer. In short, a silent observer studies the environment as it is without interaction, whereas the participant observer chronicles others' experiences of environment or even his or her own experiences and examines them for themes. Both of these techniques can provide insight into the environment studied with varying strengths and weaknesses. Numerous studies of culture using silent observation as well as ethnographic, autoethnographic, and anthropologic research exist across nonverbal communication forms, including proxemics. Yet, the impact of digital technologies on these studies is just beginning.

In early studies of proxemics, Hall favored the silent observer approach and developed a systematic notation process for studying human interactions.[14] To measure human distances, his notation system coded interactions for posture, sex, sociospatial orientation, kinesthetic factors, haptics (touch), visual field, thermal factors, olfactic space, and paralanguage, particularly volume of the voice. The notation system he describes is a sort of shorthand for coding observations of dyads in communication with one another and could be applicable across a variety of cultures, provided that someone familiar with the culture was performing the coding. Interestingly, he notes that his coding might also be well applied coding behaviors of mammals of all sorts.

Coding observations of human behaviors in spaces is an admittedly tedious process and very time consuming. The camera in its many forms provides a useful tool for researchers conducting covert or unobtrusive observations. The ability to record people moving offers the observer the ability to return to the data to justify patterns, identify precise measurements of distance, and classify participants carefully. As Hall notes in his own writing, the camera must be used unobtrusively. Photographs or video must be captured without changing the behaviors they exhibit. Marco Costa's study of interpersonal distances and group walking patterns[15] and Basch, Ethan, Zybery, and Basch's study of pedestrian behavior in intersections,[16] both described earlier, relied on cameras to capture data so that the vast information of people in motion could be slowed, examined, and studied deeply for patterns. In Costa's study, a stationary video camera was set up at a five-meter distance from a sidewalk to record groups walking in a single location of a particular sidewalk. One frame was studied for each group when the group was centered in the image. This type of stationary recording and predictable analysis allows patterns to emerge among the many groups observed, which in Costa's case numbered over 2,500 people in 1,020 separate groups.[17]

Cameras might also be effective for overt and ethnographic observations. As a recordkeeping tool, photographs and videos offer a particular vantage point for analysis. For this reason, the captured image is as much a story of the events that occurred as it is a story of the photographer who captured it. The photographer's choice of framing and when to time the selection of the image are culturally bound. Cultural cues concerning proxemics might go unnoticed and therefore unphotographed by a photographer of another culture. In cases of overt observation, Hall argues for allowing members of the culture being observed to capture their own data and to be involved in the analysis process.

This line of research offers so many applications for the study of digital proxemics. The research types—observation as data coding, ethnography, content analysis, and narrative, among others—suggest a vibrant blending of quantitative and qualitative measures that can be explored from multiple angles. At present, these types of research are spread across disciplines and types of journals and rarely intersect. Researchers studying in this interdisciplinary field of digital proxemics now have the opportunity through digital access to read each other's work and build upon it together.

Experimental and Quasi-experimental Designs

In the field of environmental psychology, some attention has been given to the use of experimental designs to study interpersonal use of space. As Jenni Harrigan reports in *The New Handbook of Methods in Nonverbal Behavior Research*,[18] projective studies were historically the most common assessment protocol in the literature. Projective studies utilize forms, miniature models, visuals, or illustrations to allow participants to mark, discuss, or otherwise indicate positions and distances in proxemics space. For example, Harrigan points to studies in which participants used felt people to indicate body positions in space, and others that used photographs of dyads or groups in which participants could identify behaviors and patterns. Psychologist John Aiello,[19] among others including Harrigan, suggests that projective studies may not translate well to real-world scenarios. Hall argued that these sorts of studies might be able to allow us to draw comparisons between cultures and explore differences by first identifying that a difference exists. Likewise, projective studies might be used to identify the presence of differences between digitized interactions and non-digitized interactions. Then, another method might be used to examine those differences more deeply.

In his meta-analysis of research on spatial behaviors, Aiello chronicles the history of the study of interpersonal distances up to the late 1980s, focusing on a variety of methods, with many ideas and constructs. Many of these approximately 700 studies he examines are more suited for research into psychological intersections with interpersonal distance or specific components of interpersonal distancing such as gaze, posture, or verbal and visual behaviors. These types of studies can provide great insight into the field of digital proxemics, particularly in understanding the impacts of technology use on interpersonal distance. Many of these types of studies, particularly surrounding gaze, are currently being explored in digital environments through usability testing. This area of research examines how people use designed products. Gaze tracking allows researchers to study website usability, but as discussed in this book, is also being applied as a sensing technology for human interface access points. Other robust usability techniques are being pioneered in the field of user experience.[20] Many of the experimental and quasi-experimental designs used to study proxemics in the past—both in environmental psychology and in usability testing—can now be replicated in present to draw conclusions about the impacts of digital technologies on human behaviors. Acting as sorts of longitudinal markers, previous studies can be compared with current studies to examine changes in behaviors over time and which changes, if any, are related to digital technologies and other instigators of cultural shift.

Cultural Critique

As Edward Hall discussed, other facets of our proxemic experiences might also be revealed through examinations and critiques of the lexicon, contemporary art, and emerging literature as markers of culture. To Hall's list of mediated markers of culture, and as an homage to Marshall McLuhan's fantastic list of media,[21] I would add media across types and modalities from the technologies that we use to the artifacts we create. Examinations of mobile device apps, video games and massively multiplayer online role playing games (MMORPGs), graphic novels, drone activities, social media messaging, and digital activism might have particular interest for critical researchers and media theorists at present. Critiques of these items offers insights into the ways that space is constructed, negotiated, and occupied. W. J. T. Mitchell's critique of the linguistic codes surrounding the 2010 Occupy movement and Arab Spring is a compelling contemporary example of this type of work.[22] Whereas Mitchell might not call his work a study in proxemics, it aptly articulates the contextual underpinnings of the use of space in cultural change and is therefore a fascinating springboard for a discussion of digital proxemics in the development of protest. These critiques may take the form of rhetorical, textual, and critical analyses from a variety of lenses.

My hope for the future of cultural critique is that the spatial turn intersecting many fields will inspire researchers to consider the spatial nature of context and location not only as a component of critique, but also as an subject to be examined thoroughly, particularly in relation to physical-digital intersections. I also enjoy and thoroughly encourage the types of popular critiques that are being broadcast by 99% Invisible,[23] published in the technology sections of Time, Wired, the New York Times, and many other outlets. I see the role of intellectual criticism as fundamentally public, and I hope that public intellectuals will continue to bring these topics of cultural shift and digital proxemics to broad audiences.

Big Data and Visualization

The rise of big data comes as no surprise in a world of ubiquitous cameras, sensors, and online communication. Big data is defined not only by the enormity of volume of data, but also by the velocity and variability of its acquisition.[24] From data capture tools, data flows rapidly into digital memory in huge amounts, but at variable peak and low times that are sometimes predictable. To these three constructs of volume, velocity, and variability, information in big data form can also be characterized by its variety and complexity.[25] In terms of variety, information is captured across many media forms, not only numeric data but also text, audio, visual, and video. This variety makes the information highly complex and requires multiple analysis techniques. Much attention continues to be given to visualization of large data sets. Generally, this is the work of translating robust numerical data into visually readable information. These are applied widely in navigation for resources we use everyday, such as traffic data on mapping application and Strava heat maps,[26] as well as to collect data in aggregate for business and marketing purposes, such as search terms, website click-throughs, and recommendation engines. In addition, locative media, including mobile phones, social media, and social networking sites, provide location-based numerical data which can be harnessed with support from a variety of applications or hacks available to partner com-panies, government agencies, and the general public in some cases. Many of the techniques listed in this chapter might be applied to investigating the remaining types of data collected, particularly data captured in visual or video formats. Big data can provide researchers with access to vast amounts of data, but translating that data into usable, meaningful information in a timely fash-ion will continue to be an onerous challenge for researchers across fields and industries. Digital proxemics, as an emerging construct in the nonverbal field, might also benefit from the advancements in big data, particularly in relation to locative media. However, researchers should endeavor to privilege the role of spatial context in analysis of such data.

Applied Research on Environmental Design

Applied research exists in a different category of work from quantitative and qualitative research designs. Applied research offers the researcher the ability to modify spaces and examine the results in real time with the goal of im-provement. This might mean improvement of the space, improvement of user experience, improvement of exhibited behaviors, or any number of other is-sues deemed improvements. As you might imagine, applied research involves the modification of an element of the environment and assessment of the results. Often, such research employs the research strategies of qualitative and quantitative measures in multifaceted assessments.

When we think about modifying spaces, we often talk about modifying the material features of the space, the work of architects, contractors, and in-terior designers. Applied research on environmental design has the potential to incorporate material and immaterial features of spaces as sites for improve-ment. To that end, I offer a rather long explanation here of a comprehensive study of space as an example of some ways we might consider the fullness of person-environment interactions through applied research. As we consider the person-environment interaction in built space, note that both the person and the environment shape each other. The built space is often a fixed fea-ture around which we must operate, but how we move is determined by both the environment and us. According to educational theorists Carney Strange and James Banning, who studied college campuses to arrive at their claims, the relationship between built spaces and humans include four components, which they articulate as physical environment, aggregate environment, orga-nizational environment, and constructed environment.[27] Together, these four components allow us to examine the ways that experiences might be designed in built spaces.

First, the physical environment refers to the design, layout, and condition of the built space. On a college campus this would include built structures and their interiors, their physical relationships to each other, the distance (or lack of distance) between buildings, the connectors used to navigate between buildings, and the collective and individual aesthetics of these structures. The physical environment includes artifacts, proxemics, distances, and navigation. Of the four components, the physical environment is the only one that references the nature of the architecture in the space. As we'll see, the other three components add human experience to the environment.

Second, the aggregate environment refers to the characteristics of people who inhabit these spaces, including aggregate data of both demographic and psychological variables. Data about the individuals in an environment generate a picture of the dominant features of that environment. In some environments, dominant features of the population are clearly and easily defined. These environments often reveal populations that are highly consistent and differentiated. For example, journalist Ron Stodghill's book *Where Everybody Looks Like Me*[28] describes the dominant features of the aggregate environments on historically black colleges and universities (HBCUs), and discusses the role of these environments in American culture. Generally, HBCUs have maintained a consistent environment through demographic homogeneity that differentiates them from the aggregate environments of other institutions by their unique sense of mission and purpose. At the collegiate level, consistency and differentiation can also be demonstrated in single-gender institutions, religious institutions, and institutions founded upon specific niche subjects, like a school of the arts or an engineering school. Environments that are both highly consistent and highly differentiated (like the examples above) allow us to label and define them, and they tend to reinforce themselves over time. Other environments which lack clear dominant features are less easy to understand or define. It is perhaps for this reason that urban centers purposefully try to brand themselves with descriptive monikers and images.

Third, the organizational environment refers to the processes and patterns that allow an environment to work toward its purposes and goals. An organizational environment exists on a spectrum balancing flexibility or rigidity based on dimensions of complexity, centralization, formalization, stratification, production, efficiency, and morale.[29] These various dimensions inform the actions and performances that are exhibited in an environment, based upon enacted rules, norms, processes, and procedures. This Weberian[30] approach to understanding an environment requires the environment and the people who control said environment to establish legitimacy—the perceived right for the organizational environment to exist. As inhabitants, we must agree that the processes and patterns of behavior in an environment are worthy of merit if we are to enact them. For example, when walking into the Sistine Chapel, visitors read signs and hear an announcement (written and spoken in multiple languages) asking them to remain quiet and not take photographs. Whether or not visitors adhere to those requests is a matter of the legitimacy they perceive in the organizational environment.

Fourth and last, constructed environments describe inhabitants' collective perceptions of the context of the culture and setting. Constructed environments are based in the framework of social construction of reality, which asks people to subjectively define their own experiences to create collective understanding.[31] Whereas the aggregate environment represents an objective look at the population as a whole, the constructed environment privileges the subjective vantage points of individuals and groups in the population. Examples of constructed environments may revolve around perceived culture, perceived communication climates, experiences of specific non-dominant populations, and perceived environmental presses. An environmental press[32] is an inhabitant's perception that his environment places a particular demand on him and the others in the environment. Examples might include a college campus that requires all students to complete internships (a practical application press), a city that bans plastic grocery bags (an environmentalist press), or a ropes course that requires four or more people to work together to complete an obstacle that cannot be mastered alone (a group welfare press). In a compelling study of wayfinding cues on the MIT campus, a group of students in a course studied revolving door usage.[33] When both are present, the use of revolving doors over swing doors conserves energy lost through air infiltration, meaning that heating and cooling requirements are reduced. The students were interested in changing the habits of people entering the building as a way to conserve energy. They found that the most effective way to increase revolving door usage was to place a large sign in front of it directing people to use the revolving door. Once the habits of revolving door use were established, signage could be removed and the behaviors persisted. In this case, wayfinding cues created an environmental press, compelling users to behave in a prescribed way by using a revolving door.

These four types of environments merge together in built spaces as we interact with them, creating a series of perceptions of and reactions to our experiences. Strange and Banning argue that one key to understanding the relationship between person and environment is to examine person-environment congruence, which they define as "the degree of fit between persons and environments."[34] Because person-environment congruence is a matter of fit, the degree of congruence between person and environment can predict how attracted an individual will be to the environment, how stable he or she will feel in the environment, and how satisfied he or she will feel as a result of being in the environment. This congruence is even more interesting in its converse, incongruence. An individual placed in an environment she perceives as incompatible will try to mitigate those feelings of incongruence. She might uproot herself from the incompatible environment to move to a new environment that she finds more congruent. If she cannot move or chooses not to move, she is likely to try to change something. Either she might try to change or alter the environment to make it more congruent with her, or she might try to change or alter herself to become more congruent with the environment. Environments alter people and people alter environments.

When digital technology enters the equation, it intersects with all four components of person-environment interactions in built space in a variety of ways. In the physical environment, technologies built into the environment become part of the design, layout, and condition of that space. In the aggregate environment, access to digital technologies and the frequency of its use likely becomes one aspect of the demography of the dominant narrative. In the organizational environment, the role of digital technologies and the processes for or barriers to its use must be negotiated. And in the constructed environments, access to digital data fundamentally changes perceptions. The issues of digital literacies emerge. These four constructs can each be studied using a variety of other methods, as is often the case in applied research. However, assessing digitized spaces using these four concepts in concert allows the researcher to view spaces with an attention to administrative concepts and user experience in the design of spaces.

A Call for Research on Digital Proxemics

Research in proxemics relies on the analysis of specific data, as well as the articulation of that data in its social and spatial context. The individual, dyadic, group, and spatial variables at play are many. As a result, research in proxe-mics can investigate perceptions of space, use of space, behaviors in space, movements through space, and the structure of space among other ideas that connect humans and spaces. As proxemics takes a digital turn, better un-derstandings of our changing culture can be explored through contemporary changes in the human use of space and the connections between physical space and digital technologies. My hope is that one or more of these methods inspires researchers to conduct their own investigations of the human use of space, and to share the findings.

APPENDIX: AN INFORMATION DESIGN PRIMER

If you want to participate in information design in the context of spaces, take a stroll down the National Mall in Washington, D.C. Walking west from the Capitol, you'll pass monuments, memorials, and buildings that tell a history of our country. Most herald the numerous successes of country, statesmen, heroes, and culture. Oversized statues of Thomas Jefferson, Abraham Lincoln, Ulysses S. Grant, and Franklin Delano Roosevelt gaze toward the Washington Monument. The likeness of Martin Luther King, Jr. looks toward the rising sun over the tidal basin on the Potomac River. The World War II memorial depicts a country of states unified around a central calling. The Smithsonian Museums lining the mall herald the success and promise of American ingenuity. Eventually, on your journey through the picturesque Constitution Gardens lining the expansive reflecting pool, you'll descend into the pit designed by Maya Lin. Her architecture will take you on a journey of discovery that is unparalleled by the other memorials in the national park. Juxtaposed against the bright, gleaming obelisk of the nearby Washington Monument, Lin's Vietnam Veterans Memorial is a glossy wall of black stone that pierces into the earth while reaching out to visitors and demanding interaction. Maya Lin's design for this monument, like that of many of her others, begs not to be viewed but rather to be experienced.

The monument is a V-shaped wall of black granite that slightly descends into the earth. Inscribed in the granite are the names of the servicemen and -women who lost their lives in the Vietnam conflict. Many friends and family members of those lost have searched for the names of their loved ones and taken rubbings of the names as mementos from the visit. However, locating a specific name to take a rubbing of the etching becomes a fairly arduous task. The approximately 58,000 names inscribed on the walls are not listed in alphabetic order but rather arranged in quadrants by the dates of their deaths. In the quest to discover the location of a particular name, the user enters into an interaction with the design of the memorial. Even viewers with no connection to the many names listed are entranced by the volume of names and the seeming disorder to the list. Lin's memorial was designed to pull every visitor into the gravity of the conflict. After experiencing the memorial, the visitor emerges back into the national park having sensed both the weight of the lives lost and the angst of the Vietnam War era in American history.

The space, quite literally, speaks of fallen servicemen and -women, but it also speaks of all of us as viewers. Walking up to the monument, one of the first things viewers notice as they face the black wall of names is their own reflection in the wall—an immediate visual interaction. Seeing one's own face reflected in the names of people who are deceased is a startling image of the temporality of life. The slight descent in elevation in the memorial is enough to cancel out some of the surrounding noise. Visitors often interact physically with the memorial as well, touching the engravings with their fingers, or taking a rubbing with a piece of paper and crayon. Many visitors even leave personal possessions like letters, clothing, flowers, or memorabilia at the memorial, using it as a site of remembrance. So many personal artifacts have been left on site that a separate archive was created to house the items that are collected daily.

At the time of its construction in 1982, the monument was highly controversial in its design. It deviated from the monuments surrounding it and was called a scar on the national mall. Thirty years later, according to *The Washington Post*,[1] it is the most visited memorial in Washington.

When Maya Lin designed this memorial, whether or not she knew exactly how it would be used, she intended it to be an experience. In her words: "I remember one of the veterans asking me before the wall was built what I thought people's reaction would be to it... I was too afraid to tell him what I was thinking, that I knew a returning veteran would cry."[2] Maya Lin understood the emotional impact of the memorial before it was finished; but her emphasis was not on the design or the designer. Rather the Vietnam Veterans Memorial emphasizes the experience of the user.

A walk southeast from the memorial leads past the storehouse museums of the Smithsonian to the United States Holocaust Memorial Museum, a sobering and emotional museum designed to be experienced. As each visitor enters the museum, he is given a new identity as one of the Jewish captives during the holocaust. Each tour group of patrons begins the tour by an initial confinement in a sterile, uninviting elevator. In my experience, the elevator ride was just enough time to discover some information about my assigned counterpart as well as to feel slightly uncomfortable about being cramped in the space. This discomfort foregrounds the experience of the museum. Throughout the walk, visitors experience glimpses of the historical context and injustice of captivity as well as sometimes gruesome remnants of the horrors of the holocaust. Users cannot exit the museum until they have walked through the entire emotional experience, finally emerging into a peaceful room intended for reflection and the recounting of the events through recent oral histories of holocaust survivors.

This emotional upheaval is no accident of design. James Ingo Freed, the museum's architect, notes that the architecture is reminiscent of Holocaust themes but not specific enough to be interpreted any one way. Instead, Freed wanted each user to interpret the architecture personally. Freed is quoted on the Museum's website as saying: "It must take you in its grip."[3] This museum is an attempt to create an experience: designed for the user from attraction through engagement to a particular conclusion.[4]

The design of experiences of space, as seen in the two memorials described above, focuses on the role of the user. This emphasis on the user's experience has shaped the contemporary practice of information design, shifting the field away from a focus on designers to a focus on users. To further investigate the design of spaces, this chapter chronicles the historical underpinnings of the field of information design and explores its trajectory toward user-experience design (UXD). In addition, placemaking is examined as one example of the role of user-experience design in the creation of places. Then, the role of mediated technologies in creating experiences is offered as a foundation for the investigation of digitized spaces through information design.

Information Design, UXD, and Placemaking

Information design specialist Beth Mazur gives us a standard and historically situated view of information design: "The field of information design applies traditional and evolving design principles to the process of translating complex, unorganized, or unstructured data into valuable, meaningful information."[5] Designers try to find the best possible way to organize (design) information so that it is the most accessible to the largest number of people. Among the early information designers were talented thinkers like William Playfair, Charles Joseph Minard, John Snow, and Florence Nightingale. Playfair invented the line, bar, and pie charts in the late 1700s. Minard, a civil engineer, integrated complex data into maps by combining geography with numerical and statistical data. Physician John Snow's map of London's cholera outbreak in 1854 helped solve a health crisis by visually revealing the relationship between cholera cases and particular water sources. Florence Nightingale, the acclaimed nurse, is famous not only because of her nursing abilities, but also because she was the first person to use information design to argue before Parliament for funding. Her rose map of the causes of mortality in the war in India led to increased funding for medicine to treat preventable disease. Like Playfair, Minard, Snow, and Nightingale, information designers use graphical and visual techniques to solve problems, bolster arguments, direct behaviors, and persuade people.

However, the process of information design is complicated by a number of factors. With the advent of reproducible media (writing via the printing press) the audience of the written medium could be expanded quickly. When media became able to depict events in temporal sequence (the telegraph and later in film), media became able to reach an audience and engage them across media-driven time. With the technological advances of broadcast media and networked media, information has become available to users anytime they desire it. These changes in the pace of media complicate the process of designing information in media forms and the ways that it can be distributed to a variety of audiences. In addition to developments in media, a recent focus on the users of designs has also complicated our understanding of information design. People in the world passively and actively receive information through various media forms. As they have been studied, the issues of usability and functionality have surfaced as determining factors of successful media.[6] But the more that media-users are studied, the more that human factors begin to play with and complicate an understanding of the ways in which users interact with designed information and information design. Aiming to understand new media coupled with a focus on human factors, contemporary media theorists and practitioners have tried to access the complexities of designing for the human condition. Thereby, the field of information design has incorporated ideas from psychology, sociology, human factors, communication, education, and other fields into its practice to better serve the user.

In 2000, Saul Carliner, a professor and past president of the Society for Technical Communication, described the trajectory of the field of technical communication in an article for the journal *Technical Communication*.[7] He asserted that the field of technical communication was previously historically dominated by the logistics and physical design elements of document design. The field of technical communication is one example of a field dominated by the historical framework focusing on the message and, thereby, the designer of the message. This trajectory incorporates fields and discourses of document design and usability testing relating to the successfulness of the design as information transmission, but does not yet tie them directly to a user's engagement with a designed object. In the article, Carliner argues for a new framework for technical communication that involves features of physical design, concepts of cognitive understanding, and issues of affective appeal.

The new framework calls designers to consider the physical, cognitive, and affective response of the user during the work of design. First, in the physical design process, the designer would consider the ways a user would find or discover presented information. This, Carliner argues, is the strength of document design practice. The second aspect of the framework, cognitive design, would incorporate the ways a user would come to understand presented information. Carliner positions this as the work of information design. Third, the affective design involves the user's desire to perform, including attention, culture, language, and motivations. Carliner offers that affective design is the most problematic of the three, and the one poised for study, terming it communication design. This three-part framework signals a need for change to match developments in media and a shift in focus from designer to user, at least in terms of technical communication. Dr. Carliner's example positions document, information, and communication design as three parts of the same goal, meeting the user and directing the response of the user to a designed end-goal.

Donald Norman, a professor and consultant in the field of user-centered design, articulated the various experiential responses of a user in his book *Emotional Design: Why We Love (or Hate) Everyday Things*.[8] Norman categorizes responses into three types: visceral, behavioral, and reflective. The visceral response is a gut reaction that hits us before we can think about it: "We either feel good or bad, relaxed or tense. Emotions are judgmental and prepare the body accordingly."[9] For Norman, cognition (e.g., the behavioral and reflective responses) comes only after the visceral response has occurred. The visceral response is so primal, so bodily, that it is pre-cognition. The gut reaction that I have to a building occurs before I can even consider whether or not to have the reaction. A behavioral response occurs during the use of a designed object. Behavioral responses deal with both form and function and the merger between the two. This response might reflect ease of use, the relationship between design and function, pleasure associated with use, and effectiveness of the design in fulfilling its purpose, the sum of which has been termed concinnity.[10] Conversely, a reflective response occurs after use. Purely cognitive, the reflective response concerns the user's memory of product and her intellectual and emotional recollection of its design. The visceral, behavioral, and reflective responses, in sum, detail the relationship between a user's intellectual and emotional responses to a design. In this framework, the user becomes vitally important. The user's responses are inseparable from her lived experience, her memory, her motivations and preferences.

In *Designing Pleasurable Products*,[11] Patrick Jordan describes four categories of pleasures experienced by users that can help to typify user-response. For Jordan, the emotional response of the user is crucial in understanding how information design functions. Functionality and usability must be combined with pleasurability to create a clear picture of information design. Jordan, a designer, consultant, and past holder of the prestigious Neirenburg Chair of Design at Carnegie-Mellon University, defines pleasure as the addition of value or the removal of need. His four forms of pleasure include ideo-pleasure, psycho-pleasure, socio-pleasure, and physio-pleasure. These pleasures arise from the emotion that comes with values, cognition, relationships, and physiological (bodily) comfort, respectively.

The works of Carliner, Norman, and Jordan all integrate the user into the information design equation. They are concerned with the ways in which a user responds to the design; but, the necessary next step in this line of thinking is an emphasis on design as a driver of user experience.[12] In *experience design 1*,[13] designer Nathan Shedroff writes that meaning resides only in the minds of the audience. Meaning equates with derived understanding: from a cognitive, behavioral, and affective response. Shedroff argues that good information design is one that takes users from data to information to knowledge to wisdom.[14] As a sometimes effective photographer, I tend to equate this process with the act of taking a photograph. The photographer sees all the data present in a scene. Using the viewfinder on the camera, he selects out the information important to display to create meaning and capture the image. In the process of being printed and archived, the print photograph becomes a container for knowledge. Then, upon viewing, the viewer can derive wisdom through a reflective and evaluative response. A photographer concerned with the user experience must take a photograph asking questions about the viewer and select information from the data that will achieve a desired response.

Shedroff argues that what is necessary for the user is an experience. The experience should attract the user, engage her in some way, and conclude the experience in a meaningful way. Much of the rest of his study in *experience design 1* is a play on response. He evokes different emotional responses throughout the study with color, design, surprise, and intrigue, all the while keeping the user in mind, allowing the user to interact with the book, and equipping the user to reflect upon and evaluate the content of the study.

The definition of information design proposed herein is grounded in Shedroff's claims surrounding experience design. It situates the user's experience throughout the process and encourages the consideration of the user, his/her experiences and environments, as well as the intended end result as designed. Our definition of information design must include composer, composed, and user. For this we turn back to Carliner: physical design, cognitive understanding, and aesthetic appeal. But Carliner's "information design" could be expanded as we consider spaces. The user is not simply processing the design in cognitive and aesthetic arenas. In addition to information processing, our definition must include the reflection of the user (an expansion of Carliner's cognitive dimension) and the final reflex of the user; that is, the user must experience the design and react to it in some way. The user's ability to accept, dismiss, or integrate the information presented in the design is of paramount importance in this definition of information design. This is an application of Shedroff's concept of information design as a transformation of data into wisdom. The user must experience the design and then respond.

For researchers in communication studies and media-related fields, this concept of user experience offers a new framework for the assessment of design in its many mediated forms, including space. Studying the experience of the user can inform practice in media production, media delivery, classroom instruction, and public speaking, to name a few. Individuals who design all forms of "documents," from oratory to televised broadcasts and from writing to Internet spaces, can consider the role that each element plays in the development of an experience for the user. Through assessing the role of the user, researchers can examine the experiences produced by the "document" and design "documents" that can engage their users in complex ways. By shifting from a design-centered focus to a user-experience focus, designers can compose "documents" in various media forms—including spaces—that offer the user more than the delivery of information.

Those who produce in this medium continually examine the style of their design to create functional spaces that continue to test the boundaries of information and experience design.

So, what is *good* design for spaces? It certainly considers the arrangement of the environment; the choice of sensory cues; the placements of interactive elements in relation to the layout of the space. But is *good design* simply characterized by beauty or by being well-conceived by the designer?

Madame L'Amic, curator of the Musée Arts Decoratifs in Paris, interviewed famed designer Charles Eames in 1969. I enjoy designer Charles Eames's response to Mme. L'Amic's question, "Is design an expression of art?" Eames says, "The design is an expression of the purpose. It may (if it is good enough) later be judged as art."[15] Building on his concept, good spatial design may be characterized as a design that fulfills the purpose for which it is designed. The role of design in an object or experience now allows us to examine not only the process of design during its creation, but also its use and its user. The examples of the memorials at the beginning of this appendix illustrate how experiences might be created in designed spaces.

One compelling application of experience design in spaces is in the context of placemaking. Ronald Lee Fleming, author of *The Art of Placemaking*, characterizes four elements of placemaking: orientation, connection, direction, and animation. "Placemaking should be the handmaiden of urban design," says Ronald Lee Fleming, author of *The Art of Placemaking*. "It should enable an urban-design strategy to vest itself through the creation of publically accessible meanings."[16] Fleming offers four design objectives that should be considered in the process of making a place: (1) spatial orientation; (2) connection; (3) direction; and (4) animation.

In the context of placemaking, spatial orientation refers to the history of a geographic location. Spatial orientation typically requires research or prior knowledge, which is often either unavailable to or hidden from the casual visitor to a geographic location. Connection refers to the semiotic cues present in the site that offer glimpses of the historical orientation. Often, these are statues, etchings, or signs from which a visitor can glean information. Direction refers to behavioral cues given to visitors about how to visit the place, where to go, and how to find what they are seeking. Direction is often handled through wayfinding practices. Finally, animation refers to the experiences and actions that emerge in a place. As people use a place, both intended and unintended behaviors start to be realized over time. Given these descriptions, some questions to be addressed in considering placemaking might include: How does a visitor act out the orientation of the place? What do people do, or how do they behave in the place? What unintended behaviors resulted from the design of the place? What improvements might be made to the place to encourage a particular behavior?

When these four elements of placemaking successfully converge, the resulting experience can be profound and deeply emotional for visitors. Consider the National 9/11 Pentagon Memorial built in honor of the 184 victims of the September 11, 2001 attack on the Pentagon in Washington, D.C. The memorial is a beautiful site which sharply demonstrates the profound power of a geographic location that is designed to call attention to its history. A quick assessment of the memorial through the lens of placemaking might illuminate the benefits of experience design that combines a geographic location, its history, and visitor behavior.

Spatial Orientation. On September 11, 2001, the hijacked American Airlines Flight 77 crashed into the Pentagon, killing 184 individuals. The victims of the attack ranged in age from 3 to 71, and though all were brought together in one horrific event in history, each had his or her own story.

Connection. According to the memorial's website, "The Pentagon Memorial captures that moment in time at 9:37 a.m. when 184 lives became intertwined for eternity. Each victim's age and location at the time of the attack have been permanently inscribed into the Memorial by the unique placement and direction of each of the 184 Memorial Units. Elegant and simple, the Pentagon Memorial serves as a timeline of the victims' ages, spanning from the youngest victim, three-year-old Dana Falkenberg, who was on board American Airlines Flight 77, to the oldest, John D. Yamnicky, 71, a Navy veteran, also aboard Flight 77 that morning."

A closer investigation of the site design reveals the element of connection. The age line, mentioned in the quote above arcs around the side of the Pentagon, and the age indicator lines flow along the same path that the plane took on its approach leading up to the crash. Each memorial unit is oriented upon this trajectory.[17] Each memorial unit is a bench which rises out of the age line and juts out over a pool of flowing water. Names are inscribed on the end of each unit. When a visitor stands at a bench and reads the name of the deceased with the Pentagon in view, the victim died in the building. Turning 180 degrees, a visitor reading a name and standing toward the open sky is recalling a victim who died aboard the airplane. In this way, the site's design compels visitors to physically orient themselves to the site's history.

In addition, the site reminds visitors that it is a somber memorial. Around the memorial units, the ground is covered in pea gravel. As a visitor walks through the memorial, the sound of gravel and the flowing water underneath the benches call for reflection. The trees are beginning to form a canopy overhead, and at night, lights shine up on the memorial units from the water underneath. These physical elements combine to remind the visitor that this location is a place of solace and remembrance.

Direction. The memorial units are arranged at the site according to age. To help visitors find a particular memorial unit, a locator stone was placed at the memorial entrance. The stone lists each victim and locates their memorial unit so that visitors may navigate the memorial. In addition, the design of the memorial is explained on the memorial's website, in signage upon entry, and on an available audio tour, ensuring that each visitor can learn about how to visit the memorial.

Animation. The Pentagon provides some general rules about behaviors in the memorial: no food or drink, no recreational activities, no animals. Visitors are encouraged to linger, to take photographs, or to leave mementos marking their visit. And, no passes or security screenings are required to visit this area of the Pentagon reservation.

But animation is more than the rules or norms of a place. Watching visitors in this place gives some clues about animation. The widow of a deceased husband might arrive and spend the afternoon reading a book on the bench memorializing her loved one. A family of visitors might single out the family of four who died aboard the plane and find each of their benches. A navy officer who lost friends in the attack may walk to a memorial unit and give a salute of honor.

Animation may also change and iterate based upon time. On September 11 of each year since the memorial's dedication, a sometimes-private, sometimes-public ceremony has been held at the memorial for families of the victims. Often, the ceremonies occur in the morning, using a moment of silence to recall the 9:37 a.m. attack. In this way, the animation element of this place also has a deep design connection to the site's orientation.

Placemaking—the successful combination of spatial orientation, connection, direction, and animation—is established through the design of the geographic location. It is designed for user experience.

Connecting Users, Spaces, and Technology through Design

Well-designed experiences are easy to catalog, especially when we are excited to participate in the design by visiting iconic landmarks, places of memory, or highly fictionalized spaces like movie sets or classic Disney properties. In a well-designed space, we embark upon adventure, losing ourselves in the experience, and wander into the spectacle of the environment surrounding us. The better the design, the less need we have to pause for directions or consult a map. We simply move.

When we encounter a well-designed experience, we get lost in it. Psychologist Mihaly Csikszentmihalyi[18] describes this experience as *flow*, the achievement of the optimal experience. One of the most common experiences of flow, as described by users, revolves around motion pictures—the idea that we can get lost in a movie. In the act of watching a movie, viewers can connect with characters on screen, emotions can be translated from the characters to the audience, and audience members can feel the emotional arc of the film through suspense, joy, fear, grief, or relief. The moments when we are fully immersed in the experience are the moments of flow.

We might enter a sense of flow when we lose ourselves in a gripping novel or feel empathy for a character in a film. Flow might be achieved in a feeling of loss when our avatar dies in a video game or a feeling of joy when a poem or painting captivates our imagination. Flow might also be achieved when we turn ourselves over to a physical experience in space. Spaces might also create experiences of flow for us when they separate us from external distraction and cause us to focus solely on the experience. I've already mentioned some built spaces that try to achieve this immersive experience: the Holocaust Museum, memorials, Main Street U.S.A. in Walt Disney World. But more extreme experiences of emotion, within the confines of a rollercoaster ride, for example, can also create experiences of flow across a shorter time period.

An experience design perspective on space considers the ways a user interacts in, around, and with space. The user feels the textures of the flooring as his feet shuffle to a point of interest. She smells the mix of building materials, air fresheners, dust, and other bodies using the space. He sees the boundaries of motion in the walls and turns his body to avoid unnecessary collisions. She analyzes the content of presented signage and makes decisions about how to orient herself. And he notices when the footsteps of others signal an intrusion into his experience of the space, or when his footsteps interrupt the experience of the space by leaving a trail of tracks behind him. Quite literally, the environment becomes an extension of the user and the user becomes part of the medium.[19] The experience of the user is as much a part of the design as the content and, as discussed above, the user must play a justified role in contemporary conceptualizations of information design.

In this context, an expanded definition of information design should include the physical design of data into an information experience; the visceral, behavioral, and cognitive responses intended for and experienced by the user; the choices made by the user; and the resulting transformation of both user and design based on the experience. This definition of information design is perhaps still too limited for the purposes of this study for one notable reason: this definition presumes that the designed object is relatively static. In our case the intersections of technology, user, and space often produce a designed space which is interactive. A digitally enhanced space both modifies the experience and responds to the actions of the user.

The addition of digital technologies to spaces (and to their users) toys with the active/passive nature of the interaction. The user and the environment both bring digital tools to the interaction, with a variety of repercussions. Smart actors and smart environments act mutually on each other. In the words of design theorists Kress and van Leeuwen, "Work transforms that which is worked on. Action changes both the actor and the environment in which and with which he or she acts."[20] The content and the action of retrieving content are important to the user. Here, Kress and van Leeuwen are discussing work and action as semiotic action; that is, the action of deriving meaning. What the study of digital and digitally enhanced spaces reminds us is that the environment is also working to derive meaning from the actor.

In the digital domain, authors, designers, and futurists have been exploring the interactivity of digital technology and users. As we might expect, these texts have largely focused on the interaction between the technology and the user to achieve the intended purpose. For example, in his book *Designing Interactions*, Bill Moggridge, designer of early laptops, asserted that designers of digital products are no longer as concerned with physical design as they are with designing the interaction between user and product. He interviewed 37 leading interaction designers to make claims about the process of designing these interactions. The text and accompanying DVD containing the interviews is a compendium of design thinking.

The purpose, therefore, in our discussion of the role of information design for spaces is to understand that digital, interactive forms of media complicate our understandings of messages and messaging. In the study of space, a turn toward contemporary theories of information design can illuminate the connection between user and the design of digitally enhanced spaces, as well as the relationship between users, digital technologies, and the spaces they inhabit.

NOTES

Preface

1. Hall, E. (1966). *The Hidden Dimension*. Anchor Books
2. Pariser, E. (2011). *The Filter Bubble: What the Internet Is Hiding from You*. New York: The Penguin Press.
3. Turkle, S. (2011). *Alone Together: Why We Expect More from Technology and Less from Each Other*. New York: Basic Books.
4. Rheingold, H. (2002). *Smart Mobs: The Next Social Revolution*. New York: Basic Books.
5. de Souza e Silva, A., & Frith, J. (2012). *Mobile Interfaces in Public Spaces*. New York: Routledge.
6. Farman, J. (2012). *Mobile Interface Theory: Embodied Space and Locative Media*. New York: Routledge.

Chapter 1 : Moving

1. Caruso, N. (2011). The Video Game Crash of 1983. *The Gaming Historian*. http://thegaminghistorian.com/the-gaming-historian-the-video-game-crash-of-1983/
2. IGN Entertainment. (2012). Top 25 Video Game Consoles of All Time. http://www.ign.com/top-25-consoles/1.html

3. See Gee, J. P. (2003). *What Video Games Have to Teach Us about Learning and Literacy*. New York: Palgrave Macmillan.

4. For a discussion of exergaming, see Bogost, I. (2010). *Persuasive Games: The expressive powers of videogames*. Cambridge, MA: MIT Press, and earlier writings, including his 2005 study on The Rhetoric of Exergaming, available at http://www.exergamefitness.com/pdf/The%20Rhetoric%20of%20Exergaming.pdf.

5. Hall, *The Hidden Dimension*, p. 1.

6. A purposeful modification of Robert Sommer's 1969 claim to teachers: "Teachers are hindered by their insensitivity to and fatalistic acceptance of the classroom environment." In Sommer, R. (1969), *Personal Space: The Behavioral Basis of Design*. Englewood Cliffs, NJ: Prentice-Hall, Inc., p. 119.

7. This type of consciousness cannot be limited to built spaces. The re-defining of initial impressions impacts every product with which I interact in any medium. Perhaps the same holds true for many consumers. The old cliché, "You can't judge a book by its cover" may be true. But we do.

8. Lindstrom, M. (2011, Nov. 7). Zones of Seduction. How Supermarkets Turn Shoppers into Hoarders. *Time Magazine*, p. 58. See also Lindstrom's book on the subject: Brandwashed: Tricks companies use to manipulate our minds and persuade us to buy (2011).

9. Read a quick report of one such outing: "I Took My Graduate Students to Walmart. Don't Judge." here: http://jamcarthur.com/2012/09/26/i-took-my-graduate-students-to-walmart-dont-judge/

10. Strange, C. C., & Banning, J. H. (2001). *Educating by Design: Creating Campus Learning Environments That Work*. San Francisco: Jossey-Bass.

11. Turkle, *Alone Together*, p. x.

12. Stenger, C. (2014). The Great Smartphone Debate, *Simply Stated* Blog, Real Simple. http://simplystated.realsimple.com/2014/07/09/the-great-smartphone-debate/

13. Jonze, S. (Producer/Writer/Director). (2013). *Her* [Motion picture]. United States: Warner Brothers Pictures.

14. Riley v. California, 573 U. S. ____ (2014).

15. Chopra, S. (2014). Computer Programs Are People Too: How Treating Smart Programs as Legal Persons Could Change Privacy as We Know It. *The Nation*. http://www.thenation.com/article/180047/computer-programs-are-people-too#

16. Pariser, *The Filter Bubble*.

17. Backstrom, L. (2013, Aug. 6). News Feed FYI: A Window into News Feed. Facebook. https://www.facebook.com/business/news/News-Feed-FYI-A-Window-Into-News-Feed

18. AFP. (2014, June 30). Facebook Under Fire for 'Creepy' Psychological Study. Yahoo! Tech. https://www.yahoo.com/tech/facebook-under-fire-for-creepy-psychological-study-90350294224.html

19. Kramer, A. D. I., Guillory, J. E., & Hancock, J. T. (2014). Experimental Evidence of Massive-Scale Emotional Contagion through Social Networks. Proceedings of the National Academy of Science. http://www.pnas.org/content/111/24/8788.full.pdf

20. Kellogg, C. (2014, June 3). Amazon and Hachette: The Dispute in 13 Easy Steps. *Los Angeles Times*. http://www.latimes.com/books/jacketcopy/la-et-jc-amazon-and-hachette-explained-20140602-story.html#page=1

21. Pariser, *The Filter Bubble*.
22. Traditional gatekeepers included editors who made decisions over the content and scope of information presented in a newspaper. While these gatekeepers still exist, digital gatekeepers are increasingly responsible for driving information, even newspaper articles, to readers.
23. Fogg, B. J. (2003). *Persuasive Technology: Using Computers to Change What We Think and Do*. San Francisco, CA: Morgan Kaufmann Publishers.
24. Hubbard, P., & Kitchin, R. (2011). *Key Thinkers on Space and Place*, 2nd Ed. Thousand Oaks, CA: Sage, p. 2.

Chapter 2 : Digitizing Proxemics

1. The anonymity of both the restaurant and the post's author call the facts of this example into question. However, the example is useful in considering the impacts of our technological extensions on business practice.
2. The debate between the use of (the correct) "restaurateur" and (the misspelled but frequently used) "restauranteur" continues among grammarians. I chose to side with the grammarians even though I like to say the wrong version with the n. Like media, language changes over time. Insert emoji here.
3. The full text of the Craigslist post, which went viral online in July 2014, was available here, but has since been removed: http://newyork.craigslist.org/mnh/rnr/4562386373.html
4. Hall, *The Hidden Dimension*.
5. This list is modified from the list offered in Julia Wood's *Communication Mosaics* textbook.
6. For example, see Preston, P. (2005). Proxemics in Clinical and Administrative Settings. *Journal of Healthcare Management*, 50(3), 151–154.
7. For example, see Crane, J., & Crane, F. G. (2010). Optimal Nonverbal Communications Strategies Physicians Should Engage in to Promote Positive Clinical Outcomes. *Health Marketing Quarterly*, 27, 262–74.
8. Smith, W., Lassiter, J., & Zee, T. (Producers), & Tennant, A. (Director). (2005). *Hitch* [Motion Picture]. United States: Sony Pictures.
9. Hall, *The Hidden Dimension*.
10. Paraphrased language from a friend's description of his experience in NYC.
11. The concept of socio-architecture is widely attributed to psychologist Humphry Osmond and architect Kyo Izumi. Osmond described sociopetal and sociofugal arrangements of space in Osmond, H. (1957). Function as the Basis of Psychiatric Ward Design. *Psychiatric Services*, 8(4), 23–29.
12. Hall, E. T. (1968). Proxemics. *Current Anthropology*, 9(2/3), 83–108.
13. Hall's article in *Current Anthropology* was published alongside seventeen responses from colleagues and reviewers in the field. These responses, along with Hall's reply, are available on pages 95–106 of the article: Hall, Proxemics.
14. Gehl, J. (2010). *Cities for People*. Washington, DC: Island Press, p. 49.
15. Montello and Sas argue that navigation can be broken down into two components: wayfinding and locomotion. See Montello, D. R., & Sas, C. (2006). Human Factors of Way-

finding in Navigation. *International Encyclopedia of Ergonomics and Human Factors*. Boca Raton, FL: CRC Press/Taylor & Francis, Ltd., pp. 2003–2008.

16. For more information on recent generational analysis, see Howe, N., & Strauss, W. (2000). *Millennials Rising: The Next Great Generation*. New York: Vintage Books, and related works.

17. See Prensky, M. (2001). Digital Natives, Digital Immigrants. *On the Horizon*, 9(5), 1–6.

18. Garvey, A. (2015). The Oregon Trail Generation: Life before and after Mainstream Tech. Retrieved from http://socialmediaweek.org/blog/2015/04/oregon-trail-generation/

19. For people born in the late 1980s and after, cassette tapes and VHS tapes were technologies on which music and movies were recorded for listening and viewing. They were (quite literally) made of printed tapes that had a knack for unspooling in the players and ruining the listening or viewing experience. Oh, and cameras used film too.

20. McLuhan, M. (1962). *The Gutenberg Galaxy: The Making of Typographic Man*. Toronto, ON: University of Toronto Press.

21. Communication technology is not the only technology or factor that influences cultural shift. Other things like changes in geography, weaponry, immigration, types of food and food sources, and relationship with water have also been positioned as movers of culture.

22. Poldma, T. (2010). Transforming Interior Spaces: Enriching Subjective Experiences through design research. *Journal of Research Practice*, 6(2), 1–12.

23. Poldma, Transforming Interior, p. 6.

24. See Poldma, T., & Wesolkowska, M. (2005). Globalisation and Changing Conceptions of Time-Space: A paradigm Shift for Interior Design? *In Form*, 5, 54–61.

25. Rheingold, H. (2002). *Smart Mobs: The Next Social Revolution*. Cambridge, MA: Basic Books, p. 84.

26. Liu, Y., Goncalves, J., Ferreira, D., Hosio, S., & Kostakos, V. (2014). Identity Crisis of Ubicomp? Mapping 15 Years of the Field's Development and Paradigm Change. Ubicomp '14 Adjunct, 75–86.

Chapter 3 : Distancing Ourselves

1. Subway cars are actually more like semi-closed environments, but they are closed enough for observations. The truly closed environments of submarines were observed by Edward Hall to shape some of his thinking on proxemics.

2. Hall, *The Hidden Dimension*.

3. Hartnett, J. J., Bailey, K. G., & Hartley, C. S. (1974). Body Height, Position, and Sex as Determinants of Personal Space. *Journal of Psychology*, 87, 129–136.

4. Caplan, M. E., & Goldman, M. (1981). Personal Space Violations as a Function of Height. *The Journal of Social Psychology*, 114(2), 167–171.

5. Remland, M. S., Jones, T. S., & Brinkman, H. (1995). Interpersonal Distance, Body Orientation, and Touch: Effects of Culture, Gender, and Age. *The Journal of Social Psychology*, 135(3), 281–297.

6. Costa, M. (2010). Interpersonal Distances in Group Walking. *Journal of Nonverbal Behavior, 34*, 15–26.
7. Knowles, E. (1972). Boundaries around Social Space: Dyadic Responses to an Invader. *Environment and Behavior, 4*(4), 437–445.
8. Knowles, E. S., Kreuser, B., Hass, S., Hyde, M, & Schuchart, G. E. (1976). Group Size and the Extension of Social Space Boundaries. *Journal of Personality and Social Psychology, 33* (5), 647–654.
9. Burgess, J. W. (1983). Interpersonal Spacing Behavior between Surrounding Nearest Neighbors Reflects Both Familiarity and Environmental Density. *Ethology and Sociobiology, 4*, 11–17; and Burgess, J. W. (1983). Developmental Trends in Proxemics Spacing between Surrounding Companions and Strangers in Casual Groups. *Journal of Nonverbal Behavior, 7*(3), 158–169.
10. Kaya, N., & Erkip, F. (1999). Invasion of Personal Space under the Condition of Short-Term Crowding: A Case Study on an Automated Teller Machine. *Journal of Environmental Psychology, 19*, 183–189.
11. Patterson, M. L., Lammers, V. M., & Tubbs, M. E. (2014). Busy Signal: Effects of Mobile Device Usage on Pedestrian Encounters. *Journal of Nonverbal Behavior, 38*(3), 313–324.
12. See the excellent chart on pp. 126–127 in Hall, *The Hidden Dimension*.
13. Self, W. (2015, January 30). "Does Technology Make People Touch Each Other Less?" A Point of View. BBC radio broadcast. http://www.bbc.com/news/magazine-31026410, transcript para. 5.
14. Roald Dahl's Willy Wonka may not have introduced the concept of Smell-o-Vision, but it was right up his alley. For a brief history of the connection between Willy Wonka and Smell-o-Vision, see http://www.seattlemet.com/articles/2010/11/15/siff-cinema-willy-wonka-smell-o-vision-1210.
15. Chang, C., Hwang, J., & Munson, S. A. (2014). Temperature Sharing to Support Remote Relationships. Ubicomp '14 Adjunct, 27–30.
16. See Ashford's website http://wearablefutures.co or Ashford, R. (2014). Responsive and Emotive Wearables: Devices, Bodies, Data and Communication. Ubicomp '14 Adjunct, 99–104.
17. Too Much Information.
18. For further review see, Hall, *The Hidden Dimension*, pp. 121–123.
19. Oswalt, P. (2014, Sept. 8–15). "Why I Quit Twitter—and Will Again." *Time Magazine*. p. 28.
20. Watson, M. D., & Atuick, E. A. (2015). Cell Phones and Alienation among Bulsa of Ghana's Upper East Region: "The Call Calls You Way." *African Studies Review, 58*(1), 113–132.
21. Watson & Atuick, Cell Phones and Alienation, p. 127.
22. Jacobson, H. (2015, April 10). The Tyranny of the Selfie. A Point of View. BBC radio broadcast. http://www.bbc.com/news/magazine-32248538
23. Jacobson, Tyranny of the Selfie, transcript para. 9.
24. https://www.facebook.com/facebook/info
25. As of Facebook's 2015 first quarter reports

26. McArthur, J. A. (2009). Digital Subculture: A geek Meaning of Style. *Journal of Communication Inquiry, 33*(1), 58–70.

27. See Hebdige, D. (1979). *Subculture: The Meaning of Style.* London: Methuen.

28. McArthur, J. A., & Bostedo-Conway, K. (2012). Exploring the Relationship between Student-Instructor Interaction on Twitter and Student Perceptions of Teacher Behaviors. *International Journal of Teaching and Learning in Higher Education, 23*(3), 286–292.

29. See for example DeGroot, J. M., Young, V. J., & VanSlette, S. (2015). Twitter Use and Its Effects on Student Perception of Instructor Credibility. *Communication Education, 64.*

30. White, A. F., & McArthur, J. A. (forthcoming). Twitter Chats as Third Places: Conceptualizing a Digital Gathering Site. *Social Media + Society.*

31. Oldenburg, R. (1999). The Great Good Place: Cafes, Coffee Shops, Bookstores, Bars, Hair Salons, and Other Hangouts at the Heart of a Community. New York: Marlowe & Co.

32. Gee, J. P. (2004). *Situated Language and Learning: A Critique of Traditional Schooling.* New York: Routledge.

33. Roschke, K. (2014). Passionate Affinity Spaces on Reddit.com: Learning Practices in the Gluten-Free Subreddit. *Journal of Digital and Media Literacy, 2*(1), Retrieved from http://www.jodml.org/2014/06/17/passionate-affinity-spaces-on-reddit-com-learning-practices-in-the-gluten-free-subreddit/

34. Gilpin, D. R. (2011). Working the Twittersphere: Microblogging as Professional Identity Construction. In Z. Papacharissi (ed.), *A Networked Self: Identity, Community, and Culture on Social Network Sites* (pp. 232–250). New York, NY: Routledge.

35. Yuqing, R., Harper, F., Drenner, S., Terveen, L., Kiesler, S., Riedl, J., & Kraut, R. E. (2012). Building Member Attachment in Online Communities: Applying Theories of Group Identity and Interpersonal Bonds. *MIS Quarterly, 36*(3), 841–864.

36. Weisgerber, C., & Butler, S (2011). Social Media as a Professional Development Tool: Using Blogs, Microblogs and Social Bookmarks to Create Personal Learning Networks. In C. Wankel (ed.), *Teaching Arts & Science with Social Media.* Bingley, U.K.: Emerald.

37. Bolter, J., & Gromala, D. (2003). *Windows and Mirrors: Interaction Design, Digital Art, and the Myth of Transparency.* Boston, MA: MIT Press.

38. Bolter & Gromala, *Windows and Mirrors,* p. 116.

39. Furr, J. (2015). The Impact of Social Media Disclosure on Persons with Physical Disabilities. Unpublished master's thesis.

40. Marquardt, N., & Greenberg, S. (2012, Jun–Aug). Informing the Design of Proxemic Interactions. *Pervasive Computing,* p. 22.

Chapter 4 : Bodies in Motion

1. See (or hear) MIT professor Dick Larson's take on queue theory in the 99% *invisible* podcast on the topic: Mars, R. (2012, March 9). Queue Theory and Design. 99% *invisible*, Ep. 49. http://99percentinvisible.org/transcript-49-queue-theory-and-design/

2. Some of these titles, including Fantasyland, Magic Kingdom, and Walt Disney World, are trademarked by Disney. For more information on Walt Disney World, visit https://disneyworld.disney.go.com/

3. For more information on Scuttle's Scavenger Hunt Interactive Displays, see http://www. insidethemagic.net/tag/scuttles-scavenger-hunt/

4. Grossman, L., & Vella, M. (2014, September 22). "Never Offline: The Apple Watch Is Just the Start. How Wearable Tech Will Change Your Life Whether You Like It or Not." *Time*, p. 44.

5. Walsh, B. (2014, Nov. 24). "Data Mine: The Next Revolution in Personal Health May Be the Little Step-Tracking Band on Your Wrist." *Time*, pp. 34–38.

6. Rochman, B. (2010, May 3). Take Two Texts and Call Me in the Morning: How a New Phone-Based Initiative Seeks to Improve Prenatal Care. *Time*, p. 54.

7. Rochman, Take Two Texts, p. 54.

8. McLuhan, M. (1964). *Understanding Media: The Extensions of Man.* Chicago: McGraw-Hill.

9. The 26 mediums (media) he addresses include media studies of the spoken word, roads and paper routes, clothing, comics, movies, weapons, ads, and clocks, among others.

10. McLuhan, *Understanding Media*, p. 8.

11. McLuhan, M. (1967). *The Medium Is the MASSAGE.* New York: Random House–Bantam Books, p. 26

12. The visual images in *The Medium Is the MASSAGE* (1967) frame this discussion in multiple ways.

13. Haraway, D. (1991). *Simians, Cyborgs, and Women: The Reinvention of Nature.* New York: Routledge, p. 164.

14. Straker, L. M., O'Sullivan, P. B., Smith, A., & Perry, M. (2007). Computer Use and Habitual Spinal Posture in Australian Adolescents. *Public Health Reports, 111*(5), pp. 634–643.

15. Young, J. G., Trudeau, M. B., Odell, D., Marinelli, K., & Dennerlein, J. T. (2013). Wrist and Shoulder Posture and Muscle Activity during Touch-screen Tablet Use: Effects of Usage Configuration, Tablet Type, and Interacting Hand. *Work, 45*(1), 59–71.

16. McAvinney, P. (2014). TEDxGreenville talk. http://tedxgreenville.com/portfolio/2014-paul-mcavinney

17. http://www.scientificamerican.com/article/next-generation-simulator/

18. Disney's *Snow White and the Seven Dwarfs* (1937)

19. Disney's *Sleeping Beauty* (1959)

20. Peter Jackson's The Lord of the Rings series of three films: The Fellowship of the Ring (2001), *The Two Towers* (2002), *The Return of the King* (2003).

21. Robert Zemekis's *The Polar Express* (2004).

22. James Cameron's *Avatar* (2009).

23. Disney/Pixar's *Ratatouille* (2007).

24. McCarthy, E. (2011, July 7). Andy Serkis & the Evolution of Performance Capture Tech, *Popular Mechanics*. Retrieved from http://www.popularmechanics.com/technology/digital/visual-effects/andy-serkis-and-the-evolution-of-performance-capture-tech

25. Peter Jackson's The Lord of the Rings series of three films.

26. McCarthy, Andy Serkis & the Evolution.

27. Anam, I., Alam, S., & Yeasin, M. (2014). Expression: A Dyadic Conversation Aid Using Google Glass for People with Visual Impairments. UbiComp '14 Adjunct, pp. 211–214.

28. Keefe, D. F. (2011). From Gesture to Form: The Evolution of Expressive Freehand Spatial Interfaces. *Leonardo, 44*(5), 460–461.

29. Rautaray, S. S., & Agrawal, A. (2015). Vision Based Hand Gesture Recognition for Human Computer Interaction: A Survey. *Artificial Intelligence Review, 43*, 1–54.
30. See http://research.microsoft.com/en-us/projects/handpose/ and http://www.fastcodesign.com/3045347/microsofts-handpose-wants-to-make-gesture-controls-something-you-actually-like
31. See Apperley. T. (2013). The Body of the Gamer: Game Art and Gestural Excess. *Digital Creativity, 24*(2), 145–156.
32. Sukale, R., Voida, S., & Koval, O. (2014). The Proxemic Web: Designing for Proxemic Interactions with Responsive Web Design. Ubicomp' 14 Adjunct, pp. 171–174.
33. McAvinney, TEDxGreenville talk.
34. Paul Verhoeven's *Robocop* (1987)
35. Jon Favreau's *Iron Man* (2008)
36. Andy Wachowski & Lana Wachowski's *The Matrix* (1999)
37. James Cameron's *The Terminator* (1984)
38. Ridley Scott's *Blade Runner* (1982)
39. Canguilhem, G. (1991). *The Normal and the Pathological.* Translated by C.R. Fawcett and R.S. Cohen. New York: Zone Books.
40. Haraway, *Simians, Cyborgs, and Women.*
41. Hayles, N. K. (1999). *How We Became Posthuman: Virtual Bodies in Cybernetics, Literature, and Informatics.* Chicago: University of Chicago Press.
42. http://www.hopkinsmedicine.org/neurology_neurosurgery/specialty_areas/epilepsy/treatment/surgery/vagus_nerve_stimulation.html
43. http://www.liberatingtech.com/
44. http://www.forbes.com/sites/andygreenberg/2012/08/13/want-an-rfid-chip-implanted-into-your-hand-heres-what-the-diy-surgery-looks-like-video/
45. See Booher, A. K. (2011). Defining Pistorius. *Disability Studies Quarterly, 31*(3), http://dsq-sds.org/article/view/1673/1598
46. www.oscarpistorius.com/about
47. Turbow, J. (2012, August 3). Bigger, Faster, Stronger. Will Bionic Limbs Put the Olympics to Shame? *Wired.* http://www.wired.com/2012/08/next-gen-prosthetics-and-sports/
48. See the 3D Systems press release here: http://www.3dsystems.com/press-releases/3d-systems-prints-first-hybrid-robotic-exoskeleton-enabling-amanda-boxtel-walk-tall
49. See http://www.amandaboxtel.com/
50. Visit the Centre for Neural Engineering online here: http://www.cfne.unimelb.edu.au/research-laboratories/bionics/
51. See the Center for Bionic Medicine online here: http://www.ric.org/research/centers/bionic-medicine/
52. DEKA Integrated Solutions in Manchester, New Hampshire. FDA approval allows DEKA to manufacture and sell the bionic arm in the US.
53. Kratochwill, L. (2015, March 16). SXSW 2015: How Extreme Bionics Will Rid the World of Disability—Robotic Limbs Will Become an Advantage, Not an Impediment. *Popular Science.* http://www.popsci.com/sxsw-2015-how-extreme-bionics-will-help-rid-world-disability
54. See the MIT press release about the powered ankle-foot prosthetic here: http://www.media.mit.edu/press/ankle/anklefoot-bg.pdf

Chapter 5 : Finding Our Way

1. Hall, *The Hidden Dimension*, p. 6.
2. Hall, *The Hidden Dimension*.
3. Castells, M. (1989). *The Informational City: Information Technology, Economic Restructuring, and the Urban Regional Process*. Oxford, UK: Blackwell.
4. Lefebvre, H. (2003). *The Urban Revolution* (R. Bononno, Trans.). Minneapolis: University of Minnesota Press. (originally published in French in 1970).
5. Gehl, *Cities for People*, pp. 6–7.
6. Gehl, *Cities for People*, p. 59.
7. One of many public plazas devoted to pedestrian and traffic controls enacted in New York City. See http://www.nyc.gov/html/dot/html/pedestrians/public-plazas.shtml
8. http://greenvillejournal.com/opinions/columns/2323-designing-downtown.html
9. Bacon, E. N. (1974). *Design of Cities* (Revised Edition). New York: Penguin Press.
10. Ito, M., Okabe, D., & Anderson, K. (2010). Portable Objects in Three Global Cities: The Personalization of Urban Places. In R. Ling & S. W. Campbell (Eds.), *The Reconstruction of Space and Time: Mobile Communication Practices*. Piscataway, NJ: Transaction Publishers.
11. Meyrowitz, J. (1985). *No Sense of Place: The Impact of Electronic Media on Social Behavior*. New York: Oxford University Press.
12. Wilken, R., & Goggin, G. (2012). *Mobile Technology and Place*. New York: Routledge.
13. Basch, C. H., Ethan, D., Zybert, P., & Basch, C. E. (2015). Pedestrian Behavior at Five Dangerous and Busy Manhattan Intersections. *Journal of Community Health*, 40, 789–792.
14. See Benyon, D., Quigley, A., O'Keefe, B., & Riva, G. (2014). Presence and Digital Tourism. *Artificial Intelligence & Society*, 29, 521–529.
15. Jain, A. (2015). Apple Inc. Experimenting with Augmented Reality. http://valuewalk.com/2015/03/apple-augmented-reality-rd-team
16. Ha, P. (2010, May 17). A Point-and-shoot Translator: Sprechen Sie Google? With a New App, Smart Phones Serve as Foreign Guides. *Time Magazine*, p. 47.
17. https://productforums.google.com/forum/#!msg/websearch/ZIiPQRDTSEQ/Yh-sXzN-36ZAJ
18. Google Translate developer update January 14, 2015, version 3.1.0
19. Rashid, Z., Pous, R., Melia-Segui, J., & Morenza-Cinos, M. (2014). Mobile Augmented Reality for Browsing Physical Spaces. Ubicomp '14 Adjunct, 155–158.
20. Rashid, Z., Pous, R., Melia-Segui, J., & Peig, E. (2014). Cricking: Browsing physical space with smart glass. Ubicomp '14 Adjunct, 151–154.
21. Chang, Y., Hou, H., Pan, C., Sung, Y., & Chang, K. (2015). Apply an Augmented Reality in Mobile Guidance to Increase Sense of Place for Heritage Places. *Educational Technology & Society*, 18(2), 166–178.
22. Fedele, A. (2015, April 15). Digital Windows Could Improve Health. http://sourceable.net/digital-windows-improve-health/#
23. See Duffy, J. F., & Czeisler, C. A. (2009). Effect of Light on Human Circadian Physiology. *Sleep Med Clin*. 4(2), 165–177.
24. http://exhibits.library.vanderbilt.edu/wordsworth.html

25. Betsworth, L., Bowen, H., Robinson, S., & Jones, M. (2014). Performative Technologies for Heritage Site Regeneration. *Personal and Ubiquitous Computing*, 18, 1631–1650.

26. The term hologram denotes its underlying holographic technological structure based in the nature of its light display. As this example, like many in current culture, uses projection screens, purists would not consider it a hologram, but a projection. Still, the way we use language changes language.

27. See the DisLife hologram in action here: https://www.youtube.com/watch?t=15&v=T-926dU9SYvg

28. Chatham, A., & Mueller, F. (2013). Adding an Interactive Display to a Public Basketball Hoop Can Motivate Players and Foster Community. UbiComp '13 Adjunct, 10pp.

29. Blake, K. S. (2001). Contested Landscapes of Navajo Sacred Mountains. *The North American Geographer, 3*(1), 29–62.

30. The San Francisco Peaks, in addition to their significance in Navajo tradition, are also the famous Kachina mountains in Hopi traditions, according to Blake, Contested Landscapes, demonstrating the differing significance of the mountains to multiple cultural groups.

31. See Farman, J. (2010) Mapping the Digital Empire: Google Earth and the Process of Postmodern Cartography. *New Media & Society, 12*, 869–888; and Farman, J. (2014). Map Interfaces and the Production of Locative Media Space. In R. Wilken, & G. Goggin (Eds.), *Locative Media.* (pp. 83–92). New York: Routledge.

32. Farman, Mapping the Digital Empire, p. 876.

33. Listen to Mars, R. (2014, March 11). One Man Is an Island. *99% Invisible* (Episode 105). Available here: http://99percentinvisible.org/episode/one-man-is-an-island/

34. Busta Rhymes is an American rapper, artist, and producer whose hit singles include "Turn It Up (Remix)/Fire It Up," "Gimme Some More," "Pass the Courvoisier, Part II," and "Touch It."

35. Farman, Mapping the Digital Empire, p. 882.

36. See Farman, *Mobile Interface Theory*, p. 50.

37. http://labs.strava.com/heatmap/#10/-77.20257/38.92927/blue/bike

38. https://www.google.com/maps/@34.8919998,-82.4277932,12z

39. http://labs.strava.com/heatmap/#12/-82.36283/34.83319/blue/bike

40. Smith, R. (2013). Mediated Competition, Comparison, and Connections: How Mobile Interactive Fitness Technologies Alter the Cycling Experience. Paper presented at the National Communication Association Annual Conference, Washington, DC.

41. de Sousa e Silva, A., & Frith, J. (2010). Locative Mobile Social Networks: Mapping Communication and Location in Urban Spaces. *Mobilities, 5*(4), 485–506.

42. Spohrer, J. (1999). Information in Places. *IBM Systems Journal, 38*(4), 602–628.

43. Butt, B. (2015). Herding by Mobile Phone: Technology, Social Network and the "Transformation" of Pastoral Herding in East Africa. *Human Ecology, 43*, 1–14.

44. http://www.latimes.com/science/sciencenow/la-sci-sn-nobel-prize-medicine-20141006-story.html

45. Read a recap of the presentations presented at the Society for Neuroscience poster session in 2010 on this topic here: http://www.abstractsonline.com/plan/ViewSession.aspx?mKey=%7bE5D5C83F-CE2D-4D71-9DD6-FC7231E090FB%7d&sKey=5e-ab0c83-2ae1-4660-b8b8-bd8c120d7199

46. Notably, apps like Waze and others have recently been employed to add user-reported data in real time to turn-by-turn navigation inside Google Maps, creating a layer of user data on top of traffic data, which is also user-based.

47. Google says its self driving car will be ready in 2020, but legal hurdles may push it back a few years: http://www.ibtimes.com/google-inc-says-self-driving-car-will-be-ready-2020-1784150

Chapter 6 : Locating Us

1. Google Street View lists its update schedule and directions for user submissions here: https://support.google.com/maps/answer/68384?hl=en

2. http://www.reddit.com/r/Portland/comments/1n323v/my_grandma_on_google_maps_street_view/

3. http://www.reddit.com/r/Portland/comments/1n323v/my_grandma_on_google_maps_street_view/cchtchd

4. Hubbard, P., & Kitchin, R. (2011). *Key Thinkers on Space and Place* (2nd ed.). Los Angeles, CA: Sage Publications.

5. See the inscription on the Newseum's Pinterest archive: https://www.pinterest.com/pin/237846424042177463/

6. http://timehop.com/

7. Fletcher, D. (2010, Sept. 6). Mayoral Runs: Where's the Craze for Location Sites Like Foursquare Taking Us? *Time Magazine*, pp. 49–50.

8. Fletcher, Mayoral Runs, p. 50.

9. Zickuhr, K. (2013, Sept. 12). Location-Based Services, Pew Research Center. http://www.pewinternet.org/2013/09/12/location-based-services/

10. de Sousa e Silva & Frith, *Mobile Interfaces in Public Spaces*.

11. Linda Henkel, professor of psychology at Fairfield University contributed through interview to the article: Wayne, T. (2015, Feb. 20). "Shutterbug Parents and Overexposed Lives." *The New York Times*. http://www.nytimes.com/2015/02/22/style/shutterbug-parents-and-overexposed-lives.html

12. Stein, J. (2012, April 23). "Don't Speak, Memory. If the Information Age Doesn't Shut Up Soon, All Our Best Stories Will Be Ruined." *Time Magazine*, p. 62.

13. See Goffman, E. (1981). *Forms of Talk*. Philadephia: University of Pennsylvania Press, and Goffman, E. (1959). *The Presentation of Self in Everyday Life*. New York: Doubleday. Also, Goffman's collective theories informing impression management are discussed in Johansson, C. (2009). On Goffman. In O. Ihlen, B. van Ruler, & M. Fredriksson (Eds.), *Public Relations and Social Theory* (pp. 119–140). New York: Routledge.

14. See Marwick, A. E., & boyd, d. (2010). I Tweet Honestly, I Tweet Passionately: Twitter Users, Context Collapse, and the Imagined Audience. *New Media & Society*, 13(1), 114–133, and Baym, N. K., & boyd, d. (2012). Socially Mediated Publicness: An Introduction. *Journal of Broadcasting & Electronic Media*, 56(3), 320–329.

15. http://www.reddit.com/r/Portland/comments/1n323v/my_grandma_on_google_maps_street_view/cchtchd

16. Lv, M., Chen, L., Chen, G., & Zhang, D. (2014). Detecting Traffic Congestions Using Cell Phone Accelerometers. UbiComp '14 Adjunct, 107–110.

17. Matsuda, Y., & Arai, I. (2014). A Safety Assessment System for Sidewalks at Night Utilizing Smartphones' Light Sensors. UbiComp '14 Adjunct, 115–118.

18. Park, K., Shin, H., & Cha, H. (2013). Smartphone-Based Pedestrian Tracking in Indoor Corridor Environments. *Perspectives on Ubiquitous Computing, 17*, 359–370.

19. Rodriguez Garzon, S., & Deva, B. (2014). Geofencing 2.0: Taking Location Based Notifications to the Next Level. Ubicomp '14 Adjunct, 921–932.

20. A synopsis of these geofence types is presented in Table 1 on page 923 in Rodriguez Garzon & Deva, Geofencing 2.0.

21. Watch iOS engineer Andrew Frederick explain beacons in Mashable's Ask a Developer series: http://mashable.com/2014/02/24/beacons-geofencing-location/

22. Kedmey, D. (2015, May 25). Caught by the Cloud: Will Better Software Help Police Put Criminals Behind Bars? *Time Magazine*, p. 18.

23. Baumann, P., Klaus, J., & Santini, S. (2014). Locator: A Self-Adaptive Framework for the Recognition of Relevant Places. UbiComp '14 Adjunct, 9–12.

24. Brown, C., Efstratiou, C., Leontiadis, I., Quercia, D., Mascolo, C., Scott, J., & Key, P. (2014). The Architecture of Innovation: Tracking Face-to-Face Interactions with Ubicomp Technologies. Ubicomp '14 Adjunct. 811–820.

25. As previously mentioned, Howard Rheingold labeled these smart rooms as one of six vectors of interaction with digitally enhanced things in Smart Mobs.

26. Liu, Y., Goncalves, J., Ferreira, D., Hosio, S., & Kostakos, V. (2014). Identity Crisis of Ubicomp? Mapping 15 Years of the Field's Development and Paradigm Change. Ubicomp '14 Adjunct, p. 84.

27. Marquardt, N., & Greenberg, S. (2012, Jun–Aug). Informing the Design of Proxemic Interactions. *Pervasive Computing*, 14–23.

28. Marquardt & Greenberg, Informing the Design of Proxemic Interactions, p. 15.

29. Matic, A., Osmani, V., & Mayora-Ibarra, O. (2012). Analysis of Social Interactions through Mobile Phones. *Mobile Network Applications, 17*, 808–819.

30. Setti, F., Russell, C., Bassetti, C., & Cristani, M. (2015). F-Formation Detection: Individuating Free-Standing Conversational Groups in Images. *PLoS ONE 10*(5), 1–26.

31. Lane, N., Pengyu, L., Zhou, L., & Zhao, F. (2014). Connecting Personal-Scale Sensing and Networked Community Behavior to Infer Human Activities. UbiComp '14 Adjunct, 595–606.

32. Wayne, "Shutterbug Parents and Overexposed Lives."

33. Monahan, T. (2015). The Right to Hide? Anti-Surveillance Camouflage and the Aestheticization of Resistance. *Communication and Critical/Cultural Studies, 12*(2), 159–178.

34. Monahan, The Right to Hide?, p. 160.

35. Madden, M., & Rainie, L. (2015, May 20). Americans' Attitudes about Privacy, Security and Surveillance. Pew Research Center Reports. http://pewinternet.org/2015/05/20/americans-attitudes-about-privacy-security-and-surveillance/

36. Stole, I. L. (2014). Persistent Pursuit of Personal Information: A Historical Perspective on Digital Advertising Strategies. *Critical Studies in Media Communication, 31*(2), 129–133. Excerpt from p. 129.

37. Pariser, *The Filter Bubble*.
38. Grossman, L. (2010, June 7). "If You Liked This . . ." *Time Magazine*, 44–48.
39. Joyner, J. (2006, Jan. 6). Dirty Apes! Walmart, Dr. King, and the Planet of the Apes. http://www.outsidethebeltway.com/dirty_apes/
40. Fletcher, Mayoral Runs, p. 50.
41. Marwick, A. (2012). The Public Domain: Social Surveillance in Everyday Life. *Surveillance & Society, 9*(4), 378–393.
42. http://www.peopleofwalmart.com/
43. Madden, & Rainie, Americans' Attitudes about Privacy.
44. These findings were reported in Rainie, L., & Madden, M. (2015, March 16). Americans' *privacy strategies post*-Snowden. Pew Research Center Reports. http://www.pewinternet.org/2015/03/16/americans-privacy-strategies-post-snowden/ and later discussed in Madden & Rainie (2015, May 20).
45. For a theoretical analysis of locative media, many of the essays address media theory and locative media in the recent edited volume: Wilken, R., & Goggin, G. (2014). *Locative Media*. New York: Routledge.
46. Riley v. California, 573 U.S. _____ (2014).
47. *Harvard Law Review* (2014, November). Riley v. California. Retrieved from http://harvardlawreview.org/2014/11/riley-v-california/
48. Rheingold, H. (2002). *Smart Mobs: The Next Social Revolution*. Cambridge, MA: Basic Books, p. 86

Chapter 7 : Inhabiting New Environments

1. Learn more about EarthCam at http://www.earthcam.com/
2. Hall, Proxemics.
3. Benjamin, W. (1968). The Work of Art in the Age of Mechanical Reproduction. In H. Arendt (Ed.), *Illuminations*. London: Fontana, p. 215.
4. Pun intended.
5. http://www.nasa.gov/mission_pages/mercury/missions/astronaut.html
6. Singh, S. (1999). *The Code Book: The Science of Secrecy from Ancient Egypt to Quantum Cryptography*. New York: Anchor Books.
7. Stafford-Frasier, Q. (2001). When Convenience Was the Mother of Invention. *Communications of the Association for Computing Machinery, 44*(7), p. 25.
8. Stafford-Frasier, Q. (1995). *The Trojan Room Coffee Pot: A (Non-technical) Biography*. Available in a series of resources including the biography of, a timeline of, and articles about the Trojan Room Coffee Pot at http://www.cl.cam.ac.uk/coffee/qsf.html
9. Koskela, H. (2003). 'Cam Era'—the Contemporary Urban Panopticon. *Surveillance and Society, 1*(3), 292–313.
10. Jimroglou, K. M. (1999). A Camera with a View: JenniCAM, Visual Representation, and Cyborg Subjectivity. *Information, Communication & Society, 2*(4), 439–453.
11. The eagles did not return to the garden as a nesting site in 2012 and have not since. For more information on the Eagle Cam, see http://norfolkbotanicalgarden.org/eagles/

12. See Katie Couric's colonoscopy on *Today*: http://www.today.com/id/3079465/t/katies-first-colonoscopy/

13. I will not debate whether Pluto is a planet or not, but it certainly looks like a planet in the New Horizons images. http://pluto.jhuapl.edu/

14. Hackman, M., & Nicas, J. (2015, July 17). Drone Delivers Medicine to Rural Viginia Clinic. *Wall Street Journal*. http://www.wsj.com/articles/drone-delivers-medicine-to-rural-virginia-clinic-1437155114

15. The National Recreation and Park Association discusses the impacts of drones on public places in Dolesh, R.J. (2015, March 1). The Drones Are Coming. http://www.parksandrecreation.org/2015/March/The-Drones-are-Coming/

16. Williams, D. (2010). The Mapping Principle, and a Research Framework for Virtual Worlds. *Communication Theory*, 20, 451–470.

17. Williams also cautions us against lumping all virtual worlds from Second Life to Worlds of Warcraft together, as they are dramatically different in purpose and use. I have done just that here, purposefully, but with apologies.

18. Hasler, B. S., & Friedman, D. (2011). Spatial Behavior in Virtual Worlds: Two Field Studies on Interavatar Distance. Paper presented at the annual meeting of the International Communication Association, Boston, MA.

19. See the collected works of Joseph Walther and progeny, and specifically: (1) Walther, J. B. (1992). Interpersonal Effects in Computer-Mediated Interaction: A Relational Perspective. *Communication Research*, 19, 52–90; (2) Walther, J. B. (1996). Computer-Mediated Communication: Impersonal, Interpersonal, and Hyperpersonal Interaction. *Communication Research*, 23, 3–43; and (3) Walther, J. (2006). Nonverbal Dynamics in Computer Mediated Communication, or :(and the Net :('s with You, :) and You :) Alone. In V. Manusov & M. Patterson (Eds.), *The Sage Handbook of Nonverbal Communication* (pp. 461–480). Thousand Oaks, CA: Sage.

20. More to come on digital literacies in the next chapter.

21. Steuer, J. (1992). Defining Virtual Reality: Dimensions Determining Telepresence. *Journal of Communication*, 4, 73–93.

22. Kedmey, D. (2015, Feb. 9). Virtually Real: Microsoft Joins the Crowd Betting on 3-D Headsets. *Time Magazine*, p. 12.

23. https://www.oculus.com/en-us/blog/the-oculus-rift-oculus-touch-and-vr-games-at-e3/

24. Jeep Barnett and many other designers were interviewed by Joel Stein for a *Time Magazine* cover story on virtual reality. See Stein, J. (2015, Aug. 17). Inside the Box. *Time Magazine*, pp. 40–49.

25. Ring, B. (2015, Aug. 13). Zero Latency: The VR revolution begins in Melbourne, Australia. CNET. http://www.cnet.com/news/zero-latency-vr-entertainment-revolution-begins-melbourne-australia/

26. Danielson, T. (2015, May 12). Virtual Reality Theme Park Blurs Lines between Digital and Physical (+ Video). *The Christian Science Monitor*. http://www.csmonitor.com/Technology/2015/0512/Virtual-reality-theme-park-blurs-lines-between-digital-and-physical-video

27. Baudrillard, J. (Glaser, S. F., trans.). (1994). Simulacra and Simulation. Ann Arbor: University of Michigan Press.

Chapter 8 : Developing Literacies

1. I recognize that even this language helps me make the point I am about to make. "Swipe" in my daughter's terms would mean to move her finger across the glass whereas "swipe" in her grandmother's terms would mean to steal. My daughter was trying to do the former, not the latter.
2. Prensky, M. (2001). Digital Natives, Digital Immigrants. *On the Horizon*, 9(5), 1–6.
3. If you haven't listened to the 99% *Invisible* podcast on UTBAPHs (repurposed buildings that "Used To Be A Pizza Hut"), it is worth the time spent: Mars, R. (2014, Feb. 25). U.T.B.A.P.H. 99% *Invisible*, 103. http://99percentinvisible.org/episode/u-t-b-a-p-h/
4. The term "sentient objects" is borrowed from Howard Rheingold's *Smart Mobs*, Chapter 4.
5. Vermesan, O., & Friess, P. (Eds.). (2014). *Internet of Things—From Research and Innovation to Market Deployment*. Aalborg, DK: River Publishers.
6. In 2015, Google restructured into Alphabet, an umbrella group with oversight of Google, Nest, the self-driving car initiative, and a baker's dozen other subsidiaries. See Alphabet CEO Larry Page's letter to investors: Page, L. (2015, Aug. 10). Google Announces Plan for New Operating Structure. https://investor.google.com/releases/2015/0810.html
7. Perera, C., Liu, H., & Jayawardena, S. (2015). The Emerging Internet of Things Marketplace from an Industrial Perspective: A Survey. Emerging Topics in Computing, IEEE Transactions. Available in pre-print: http://ieeexplore.ieee.org/xpl/articleDetails.jsp?reload=true&arnumber=7004800
8. The James L. Knight School of Communication at Queens University of Charlotte was endowed with a grant from Knight Foundation to move the needle on digital and media literacy in the city of Charlotte. I was part of the team that developed, authored, and oversaw the grant in its early days. Even though this book is not sponsored by our grant initiatives, the grant process and its focus on digital literacies taught me a great many things, some of which I share herein.
9. Gee, J. P. (2003). *What Video Games Have to Teach Us about Learning and Literacy*. New York: Palgrave Macmillan, p. 20.
10. This section was recounted in an earlier form on my website: http://jamcarthur.com/2010/10/16/defining-digital-and-media-literacy/
11. Hobbs, R. (2010). *Digital and Media Literacy: A Plan of Action*. Washington, DC: Aspen Institute. p. vii
12. The *Journal of Digital and Media Literacy* is available online at http://jodml.org
13. Rheingold, H. (2010). Attention, and Other 21st Century Social Media Literacies. *EDUCAUSE Review*, 45(5), 14–24.
14. Sommer, R. (1969) *Personal Space: The Behavioral Basis of Design*. Englewood Cliffs, NJ: Prentice-Hall, Inc., p. 119.
15. These three levels were theorized by Donald A. Norman in *Emotional Design: Why We Love (or Hate) Everyday Things* and were introduced in chapter 2. Here, I am applying them in the specific context of spaces.
16. Poeschel, S., & Doring, N. (2010). Nonverbal Cues in Mobile Phone Text Messages: The Effects of Chronemics and Proxemics. In R. Ling & S. Campbell (Eds.), *The Reconstruc-*

tion of Space and Time: Mobile Communication Practices. New Brunswick, NJ: Transaction Publishers.

17. See chapter 7, the subsection titled "We archive us."
18. Farman, *Mobile Interface Theory*, p. 17.
19. de Sousa e Silva, A. (2006). Mobile Technologies as Interfaces of Hybrid Spaces. *Space and Culture*, 9(3), 261–278. Quote from p. 269.
20. Thanks to Adrienne Burris for sharing this post, and for agreeing to include it here. Original post here: https://www.facebook.com/adrienne.k.rankin/posts/10102145962470928
21. David A. Gruenewald now lists his name on publications as David A. Greenwood.
22. Gruenewald, D. A. (2003). The Best of Both Worlds: A Critical Pedagogy of Place. *Educational Researcher*, *32*(4), 3–12.
23. Arieff, A. (2015, Sept. 6). The Internet of Way Too Many Things. *New York Times.* http://nytimes.com/2015/09/06/opinion/sunday/allison-arieff-the-internet-of-way-too-many-things.html
24. http://shop.leeo.com/
25. http://www.whistle.com/tagg-gps-pet-tracker/
26. http://mimobaby.com/
27. Mattern, S. (2014, April). Interfacing Urban Intelligence. *Places Journal.* https://places-journal.org/article/interfacing-urban-intelligence/ p. 3 of 33.
28. Baym & boyd, Socially Mediated Publicness, p. 328.
29. Golumbia, D. (2008). Computers and Cultural Studies. In R. Kolker (Ed.), *The Oxford Handbook of Film and Media Studies* (pp. 508–526). New York: Oxford University Press.
30. Newitz, A. (2015, April 1). The Guerilla Movement to Deploy Sensors All Over Your City. GrayArea.org (Gray Area Foundation for the Arts). http://grayarea.org/press/the-guerrilla-movement-to-deploy-sensors-all-over-your-city/
31. Johnson, S. (2010, May 31). In Praise of Oversharing. *Time Magazine*, p. 39.
32. Sarasohn-Kahn, J. (2008). *The Wisdom of Patients: Healthcare meets online social media.* Oakland: California HealthCare Foundation, p. 8.
33. Fox, S., & Purcell, K. (2010, March 24). Chronic Disease and the Internet. Pew Internet and American Life Project. http://www.pewinternet.org/files/old-media//Files/Reports/2010/PIP_Chronic_Disease_with_topline.pdf
34. Mitchell, W.J.T. (2012). Image, Space, Revolution: The Arts of Occupation. *Critical Inquiry, 39*(1), 8–32.
35. 115,000,000 users (and growing) as of September 2015, according to https://www.change.org/about
36. For more information, visit http://change.org or read Ghosh, B., & Dias, E. (2012, April 9). Change Agent: Ben Rattray's Petition Website Allows Anyone to Turn a Complaint into a Cause—and Win. *Time Magazine*, pp. 40–43.
37. See Goffman (1983), Blake (2001), Gruenewald (2003), Gehl (2010), Bax (2011), and Mitchell (2012). This small sampling of sources underlies the wide diversity of spaces that can include this debate merging interpersonal rituals, religion, education, architecture, political landscape, and cultural practice. Many more could be referenced here.
38. See George Orwell's *1984*, Chapter 4, or hear me read from it here: https://www.youtube.com/watch?v=vzEVnPMJifQ

Chapter 9 : Researching Digital Proxemics

1. Hall, Proxemics.
2. Included in the published commentary in Hall, Proxemics, p. 97.
3. Included in the published commentary in Hall, Proxemics, p. 95
4. Included in the published commentary in Hall, Proxemics, p. 96.
5. Included in the published commentary in Hall, Proxemics, p. 96.
6. Included in the published commentary in Hall, Proxemics, p. 98.
7. Included in the published commentary in Hall, Proxemics, p. 103.
8. Published author reply to comments included in Hall, Proxemics, p. 106.
9. My hope is that this list will be regarded in the same way: inspiring and incomplete.
10. Hall, E. T. (1974). *Handbook for Proxemic Research.* Washington, DC: Society for the Anthropology of Visual Communication.
11. The closest one to me was at Davidson University's library. As it was published by a society rather than a publisher, its circulation was limited, resulting in few available copies.
12. The World catalog of the Online Computer Library Center (OCLC): https://www.oclc.org/worldcat.en.html
13. I purposefully avoided the term "normal body" here.
14. Hall, E. T. (1963). A System for the Notation of Proxemics Behavior. *American Anthropologist*, 65(5), 1003–1026.
15. Dicussed in Chapter 5. Costa, M. (2010). Interpersonal Distances in Group Walking. *Journal of Nonverbal Behavior, 34,* 15–26.
16. Discussed in Chapter 6. Basch, C. H., Ethan, D., Zybert, P., & Basch, C. E. (2015). Pedestrian Behavior at Five Dangerous and Busy Manhattan Intersections. *Journal of Community Health, 40,* 789–792.
17. Costas, M. (2010). p. 18.
18. Harrigan, J. A. (2008). Proxemics, Kinesics, and Gaze. In J. A. Harrigan, R. Rosenthal, & K. R. Scherer (Eds.), *The New Handbook of Methods in Nonverbal Behavior Research* (pp. 137–198). New York: Oxford University Press.
19. Aiello, J. R. (1987). Human Spatial Behavior. In D. Stokols, & I. Altman (Eds.). *Handbook of Environmental Psychology* (pp. 389–504). New York: John Wiley & Sons.
20. One excellent resource for joining this conversation is http://UXMatters.com, a site which chronicles developments in the UX and usability testing field.
21. McLuhan, *Understanding Media.*
22. Mitchell, Image, Space, Revolution.
23. http://99percentinvisible.org
24. Laney, D. (2001, Feb. 6). 3D Data Management: Controlling Data Volume, Velocity, and Variety. META Group Application Delivery Strategies. Available online at http://blogs.gartner.com/doug-laney/files/2012/01/ad949-3D-Data-Management-Controlling-Data-Volume-Velocity-and-Variety.pdf
25. Troester, M. (2012). Big Data Meets Big Data Analytics [White Paper]. Cary, NC: SAS Institute, Inc. Retrieved from http://www.sas.com/content/dam/SAS/en_us/doc/whitepaper1/big-data-meets-big-data-analytics-105777.pdf

26. http://www.strava.com

27. Strange, C. C., & Banning, J. H. (2001). *Educating by Design: Creating Campus Learning Environments That Work*. San Francisco: Jossey-Bass, p. 5.

28. Stodghill, R. (2015). *Where Everybody Looks Like Me: At the Crossroads of America's Black Colleges and Culture*. New York: Harper Collins.

29. This list of dimensions are the subtitles of Strange and Banning's (*Educating by Design*) chapter on organizational environments (pp. 57–82).

30. Max Weber's approach to organizing lays the groundwork for this understanding of environments, particularly surrounding the ways that an organization builds legitimacy among its citizens. See Wæraas, A. (2009). On Weber. In Ihlen, Ø., van Ruler, B., & Fredriksson, M. *Public Relations and Social Theory: Key Figures and Concepts* (pp. 301–322). New York: Routledge.

31. See Berger, P. L., & Luckmann, T. (1966). *The Social Construction of Reality: A Treatise in the Sociology of Knowledge*. Garden City, NY: Doubleday., and subsidiary or supplementary works.

32. Strange and Banning, *Educating by Design* adopted the concept of environmental press from two previous works which are listed here for further review:
    (1) Pace, C. R., & Stern, G. G. (1958). An Approach to the Measurement of Psychological Characteristics of College Environments. *Journal of Educational Psychology, 49*, 269–277.
    (2) Walsh, W. B. (1973). *Theories of Person-Environment Interaction: Implications for the College Student*. Iowa City, IA: American College Testing Program.

33. Cullum, B. A., Lee, O., Sukkasi, S., & Weslowski, D. (2006). Modifying Habits towards Sustainability: A Study of Revolving Door Usage on the MIT Campus. http://web.mit.edu/~slanou/www/shared_documents/366_06_REVOLVING_DOOR.pdf

34. Strange & Banning, *Educating by Design*, p. 52.

Appendix : An Information Design Primer

1. http://www.washingtonpost.com/gog/museums/vietnam-veterans-memorial,798075.html

2. Lin, M. (2000). *Boundaries*. New York: Simon & Schuster, p. 4:16

3. United States Holocaust Memorial Museum (2006). Unites States Holocaust Memorial Museum. Retrieved December 15, 2006, from http://www.ushmm.org

4. See Shedroff, N. (2001). *Experience Design 1*. Indianapolis, IN: New Riders Publishing, p. 4.

5. Albers, M., & Mazur, B. (Eds.) (2003). *Content and Complexity: Information Design in Technical Communication*. Mahwah, NJ: Lawrence Erlbaum & Associates, Inc., p. 23.

6. See Albers & Mazur, *Content and Complexity*.

7. Carliner, S. (2000). Physical, Cognitive, & Affective: A Three-Part Framework for Information Design. *Technical Communication*. 561–576.

8. Norman, *Emotional Design*.

9. Norman, *Emotional Design*, p. 13.

10. Concinnity is a term borrowed from Coates, D. (2003). *Watches Tell More Than Time: Product Design, Information, and the Quest for Elegance*. New York: McGraw-Hill.

11. Jordan, P. A. (2000). *Designing Pleasurable Products*. London: Taylor & Francis, Inc.

12. This study investigates the applications of these ideas to spaces. For an overview of the ways this might direct a conversation within the field of technical communication, see Williams, S. (2007). User Experience Design for Technical Communication: Expanding Our Notions of Quality Information Design. *Proceedings of the Annual Meeting of the IEEE Professional Communication Society*.

13. Shedroff, *Experience Design 1*.

14. Shedroff, *Experience Design 1*, p. 34

15. http://www.brainpickings.org/index.php/2011/10/03/charles-eames-on-design-1972/

16. Fleming, R. L. (2007). *The Art of Placemaking*. New York: Merrell, p. 19

17. View a layout of the Pentagon Memorial through the memorial's interactive map, available at http://www.pentagonmemorial.net/explore/interactive-map

18. Csikszentmihalyi, M. (1990). *Flow: The Psychology of Optimal Experience*. New York: Harper & Row Publishers.

19. McLuhan, *The Medium is the MASSAGE*.

20. Kress, G., & van Leeuwen, T. (2001). *Multimodal Discourse: The Modes and Media of Contemporary Communication*. London: Hodder-Arnold, p. 36.

REFERENCES

AFP. (2014, June 30). Facebook Under Fire for "Creepy" Psychological Study. *Yahoo! Tech.* https://www.yahoo.com/tech/facebook-under-fire-for-creepy-psychological-study-90350294224.html

Aiello, J. R. (1987). Human Spatial Behavior. In D. Stokols, & I. Altman. *Handbook of Environmental Psychology* (pp. 389–504). New York: John Wiley & Sons.

Albers, M., & Mazur, B. (Eds.) (2003). *Content and Complexity: Information Design in Technical Communication.* Mahwah, NJ: Lawrence Erlbaum & Associates, Inc.

Anam, I., Alam, S., & Yeasin, M. (2014). Expression: A Dyadic Conversation Aid Using Google Glass for People with Visual Impairments. *UbiComp '14 Adjunct,* 211–214.

Ashford, R. (2014). Responsive and Emotive Wearables: Devices, Bodies, Data And Communication. *Ubicomp '14 Adjunct,* 99–104.

Apperley, T. (2013). The Body of the Gamer: Game Art and Gestural Excess. *Digital Creativity,* 24(2), 145–156.

Arieff, A. (2015, Sept. 6). The Internet of Way Too Many Things. *New York Times.* http://nytimes.com/2015/09/06/opinion/sunday/allison-arieff-the-internet-of-way-too-many-things.html

Backstrom, L. (2013, Aug. 6). News Feed FYI: A Window into News Feed. Facebook. https://www.facebook.com/business/news/News-Feed-FYI-A-Window-Into-News-Feed

Bacon, E. N. (1974). *Design of Cities (Revised Edition).* New York: Penguin Press.

Basch, C. H., Ethan, D., Zybert, P., & Basch, C. E. (2015). Pedestrian Behavior at Five Dangerous and Busy Manhattan Intersections. *Journal of Community Health,* 40, 789–792.

Baudrillard, J. (Glaser, S. F., trans.). (1994). *Simulacra and Simulation*. Ann Arbor, MI: University of Michigan Press.

Baumann, P., Klaus, J., Santini, S. (2014). Locator: A Self-Adaptive Framework for the Recognition of Relevant Places. *UbiComp '14 Adjunct*, 9–12.

Bax, M. (2011). An Evolutionary Take on (Im)Politeness: Three Broad Cevelopments in the Marking out of Socio-Proxemic space. *Journal of Historical Pragmatics, 12*(1-2), 255-282.

Baym, N. K., & boyd, d. (2012). Socially Mediated Publicness: An Introduction. *Journal of Broadcasting & Electronic Media, 56*(3), 320–329.

Benjamin, W. (1968). The Work of Art in the Age of Mechanical Reproduction. In H. Arendt (Ed.), *Illuminations*, (pp. 214–218). London: Fontana.

Benyon, D., Quigley, A., O'Keefe, B., & Riva, G. (2014). Presence and Digital Tourism. *Artificial Intelligence & Society, 29*, 521–529.

Berger, P. L., & Luckmann, T. (1966). *The Social Construction of Reality: A Treatise in the Sociology of Knowledge*. Garden City, NY: Doubleday.

Betsworth, L., Bowen, H., Robinson, S., & Jones, M. (2014). Performative Technologies for Heritage Site Regeneration. *Personal and Ubiquitous Computing, 18*, 1631–1650.

Blake, K. S. (2001). Contested Landscapes of Navajo Sacred Mountains. *The North American Geographer, 3*(1), 29–62.

Bogost, I. (2010). *Persuasive Games: The Expressive Powers of Videogames*. Cambridge, MA: MIT Press.

Bolter, J., & Gromala, D. (2003). *Windows and Mirrors: Interaction Design, Digital Art, and the Myth of Transparency*. Cambridge, MA: MIT Press.

Booher, A. K. (2011). Defining Pistorius. *Disability Studies Quarterly, 31*(3), http://dsq-sds.org/article/view/1673/1598

Brown, C., Efstratiou, C., Leontiadis, I., Quercia, D., Mascolo, C., Scott, J., & Key, P. (2014). The Architecture of Innovation: Tracking Face-to-Face Interactions with Ubicomp Technologies. *Ubicomp '14 Adjunct*. 811–820.

Burgess, J. W. (1983). Interpersonal Spacing behavior between Surrounding Nearest Neighbors Reflects Both Familiarity and Environmental Density. *Ethology and Sociobiology, 4*, 11–17.

Burgess, J. W. (1983). Developmental trends in Proxemics Spacing between Surrounding Companions and Strangers in Casual Groups. *Journal of Nonverbal Behavior, 7*(3), 158–169.

Butt, B. (2015). Herding by Mobile Phone: Technology, Social Network and the "Transformation" of Pastoral Herding in East Africa. *Human Ecology, 43*, 1–14.

Canguilhem, G. (1991). *The Normal and the Pathological*. Translated by C. R. Fawcett and R. S. Cohen. New York: Zone Books.

Caplan, M. E., & Goldman, M. (1981). Personal Space Violations as a Function of Height. *The Journal of Social Psychology, 114*(2), 167–171.

Carliner, S. (2000). Physical, Cognitive, and Affective: A Three-Part Framework for Information Design. *Technical Communication*, 561–576.

Caruso, N. (2011). The Video Game Crash of 1983. The Gaming Historian. http://thegaming-historian.com/the-gaming-historian-the-video-game-crash-of-1983/

Castells, M. (1989). *The Informational City: Information Technology, Economic Restructuring, and the Urban Regional Process*. Oxford, UK: Blackwell.

Chang, Y., Hou, H., Pan, C., Sung, Y., & Chang, K. (2015). Apply an Augmented Reality in Mobile Guidance to Increase Sense of Place for Heritage Places. *Educational Technology & Society, 18*(2), 166–178.

Chang, C., Hwang, J., & Munson, S. A. (2014). Temperature Sharing to Support Remote Relationships. *Ubicomp '14 Adjunct*, 27–30.

Chatham, A., & Mueller, F. (2013). Adding an Interactive Display to a Public Basketball Hoop Can Motivate Players and Foster Community. *UbiComp '13 Adjunct*.

Chopra, S. (2014). Computer Programs Are People Too: How Treating Smart Programs as Legal Persons Could Change Privacy as We Know It. *The Nation*. http://www.thenation.com/article/180047/computer-programs-are-people-too#

Coates, D. (2003). *Watches Tell More Than Time: Product Design, Information, and the Quest for Elegance*. New York: McGraw-Hill.

Costa, M. (2010). Interpersonal Distances in Group Walking. *Journal of Nonverbal Behavior, 34*, 15–26.

Crane, J., & Crane, F. G. (2010). Optimal Nonverbal Communications Strategies Physicians Should Engage in to Promote Positive Clinical Outcomes. *Health Marketing Quarterly, 27*, 262–274.

Cullum, B. A., Lee, O., Sukkasi, S., & Weslowski, D. (2006). *Modifying Habits towards Sustainability: A Study of Revolving Door Usage on the MIT Campus*. http://web.mit.edu/~slanou/www/shared_documents/366_06_REVOLVING_DOOR.pdf

Csikszentmihalyi, M. (1990). *Flow: The Psychology of Optimal Experience*. New York: Harper & Row Publishers.

Danielson, T. (2015, May 12). Virtual Reality Theme Park Blurs Lines between Digital and Physical (+Video). *The Christian Science Monitor*. http://www.csmonitor.com/Technology/2015/0512/Virtual-reality-theme-park-blurs-lines-between-digital-and-physical-video

de Sousa e Silva, A. (2006). Mobile Technologies as Interfaces of Hybrid Spaces. *Space and Culture, 9*(3), 261–278.

de Sousa e Silva, A., & Frith, J. (2010). Locative Mobile Social Networks: Mapping Communication and Location in Urban Spaces. *Mobilities, 5*(4), 485–506.

de Sousa e Silva, A., & Frith, J. (2012). *Mobile Interfaces in Public Spaces: Locational Privacy, Control, and Urban Sociability*. New York: Routledge.

DeGroot, J. M., Young, V. J., & VanSlette, S. (2015). Twitter Use and Its Effects on Student Perception of Instructor Credibility. *Communication Education, 64*.

Dolesh, R.J. (2015, March 1). The Drones Are Coming. http://www.parksandrecreation.org/2015/March/The-Drones-are-Coming/

Duffy, J. F., & Czeisler, C. A. (2009). Effect of Light on Human Circadian Physiology. *Sleep Med Clin. 4*(2), 165–177.

Farman, J. (2010) Mapping the Digital Empire: Google Earth and the Process of Postmodern Cartography. *New Media & Society, 12*, 869–888.

Farman, J. (2012). *Mobile Interface Theory: Embodied Space and Locative Media*. New York: Routledge.

Farman, J. (2014). Map Interfaces and the Production of Locative Media Space. In R. Wilken, & G. Goggin (Eds.), *Locative Media* (pp. 83–92). New York: Routledge

Fedele, A. (2015, April 15). Digital Windows Could Improve Health. http://sourceable.net/digital-windows-improve-health/#

Fleming, R.L. (2007). *The Art of Placemaking*. New York: Merrell.

Fletcher, D. (2010, Sept. 6). Mayoral Runs: Where's the Craze for Location Sites Like Foursquare Taking Us? *Time Magazine*, pp. 49–50.

Fogg, B. J. (2003). *Persuasive Technology: Using Computers to Change What We Think and Do*. San Francisco, CA: Morgan Kaufmann Publishers.

Fox, S., & Purcell, K. (2010, March 24). Chronic Disease and the Internet. Pew Internet and American Life Project. http://www.pewinternet.org/files/old-media//Files/Reports/2010/PIP_Chronic_Disease_with_topline.pdf

Furr, J. (2015). The Impact of Social Media Disclosure on Persons with Physical Disabilities. Unpublished master's thesis.

Garvey, A. (2015). The Oregon Trail Generation: Life before and after Mainstream Tech. Retrieved from http://socialmediaweek.org/blog/2015/04/oregon-trail-generation/

Gee, J. P. (2003). *What Video Games Have to Teach Us about Learning and Literacy*. New York: Palgrave Macmillan.

Gee, J. P. (2004). *Situated Language and Learning: A Critique of Traditional Schooling*. New York: Routledge.

Gehl, J. (2010). *Cities for People*. Washington, DC: Island Press.

Ghosh, B., & Dias, E. (2012, April 9). Change Agent: Ben Rattray's Petition Website Allows Anyone to Turn a Complaint into a Cause—and Win. *Time Magazine*. pp. 40–43.

Gilpin, D. R. (2011). Working the Twittersphere: Microblogging as Professional Identity Construction. In Z. Papacharissi (ed), *A Networked Self: Identity, Community, and Culture on Social Network Sites* (pp. 232–250). New York, NY: Routledge.

Goffman, E. (1959). *The Presentation of Self in Everyday Life*. New York, NY: Doubleday.

Goffman, E. (1981). *Forms of talk*. Philadelphia, PA: University of Pennsylvania.

Golumbia, D. (2008). Computers and Cultural Studies. In R. Kolker (Ed.). *The Oxford Handbook of Film and Media Studies* (pp. 508–526). New York: Oxford University Press.

Grossman, L. (2010, June 7). "If You Liked This..." *Time Magazine*, 44–48.

Grossman, L., & Vella, M. (2014, September 22). "Never Offline: The Apple Watch Is Just the Start. How Wearable Tech Will Change Your Life Whether You Like It or Not." *Time*. p. 44.

Gruenewald, D.A. (2003). The Best of Both Worlds: A Critical Pedagogy of Place. *Educational Researcher, 32*(4), 3–12.

Hackman, M., & Nicas, J. (2015, July 17). Drone Delivers Medicine to Rural Virginia Clinic. *Wall Street Journal*. http://www.wsj.com/articles/drone-delivers-medicine-to-rural-virginia-clinic-1437155114

Hall, E. T. (1963). A System for the Notation of Proxemics Behavior. *American Anthropologist, 65*(5), 1003–1026.

Hall, E. T. (1966). *The Hidden Dimension*. Garden City, NY: Anchor Books.

Hall, E. T. (1968). Proxemics. *Current Anthropology, 9*(2/3), 83–108.

Hall, E. T. (1974). *Handbook for Proxemic Research*. Washington, DC: Society for the Anthropology of Visual Communication.

Haraway, D. (1991). *Simians, Cyborgs, and Women: The Reinvention of Nature*. New York: Routledge.

Harrigan, J. A. (2008). Proxemics, Kinesics, and Gaze. In J. A. Harrigan, R. Rosenthal, & K. R. Scherer. *The New Handbook of Methods in Nonverbal Behavior Research* (pp. 137–198). New York: Oxford University Press.

Hartnett, J. J., Bailey, K. G., & Hartley, C. S. (1974). Body Height, Position, and Sex as Determinants of Personal Space. *Journal of Psychology*, 87, 129–136.

Harvard Law Review. (2014, November). *Riley V. California*. Retrieved from http://harvardlaw-review.org/2014/11/riley-v-california/

Hasler, B. S., & Friedman, D. (2011). Spatial Behavior in Virtual Worlds: Two Field Studies on Interavatar Distance. Paper presented at the annual meeting of the International Communication Association, Boston, MA.

Hayles, N. K. (1999). *How We Became Posthuman: Virtual Bodies in Cybernetics, Literature, and Informatics*. Chicago: University of Chicago Press.

Hebdige, D. (1979). *Subculture: The Meaning of Style*. London: Methuen.

Hobbs, R. (2010). *Digital and Media Literacy: A Plan of Action*. Washington, DC: Aspen Institute.

Howe, N., & Strauss, W. (2000). *Millennials Rising: The Next Great Generation*. New York: Vintage Books.

Hubbard, P., & Kitchin, R. (2011). *Key Thinkers on Space and Place, 2nd Ed*. Thousand Oaks, CA: Sage.

IGN Entertainment (2012). Top 25 Video Game Consoles of All Time. http://www.ign.com/top-25-consoles/1.html

Ito, M., Okabe, D., & Anderson, K. (2010). Portable Objects in Three Global Cities: The Personalization of Urban Places. In R. Ling & S. W. Campbell (Eds.). *The Reconstruction of Space and Time: Mobile Communication Practices*. Piscataway, NJ: Transaction Publishers.

Jacobson, H. (2015, April 10). "The Tyranny of the Selfie." *A Point of View*. BBC radio broadcast. http://www.bbc.com/news/magazine-32248538.

Jain, A. (2015). Apple Inc. Experimenting with Augmented Reality. http://valuewalk.com/2015/03/apple-augmented-reality-rd-team

Jimroglou, K. M. (1999). A Camera with a View: JenniCAM, Visual Representation, and Cyborg Subjectivity. *Information, Communication & Society*, 2(4), 439–453.

Johansson, C. (2009). On Goffman. In O. Ihlen, B. van Ruler, & M. Fredriksson (Eds.), *Public Relations and Social Theory* (pp. 119–140). New York: Routledge.

Johnson, S. (2010, May 31). In Praise of Oversharing. *Time Magazine*, p. 39.

Jordan, P. A. (2000). *Designing Pleasurable Products*. London: Taylor & Francis, Inc.

Joyner, J. (2006, Jan. 6). Dirty Apes! Walmart, Dr. King, and the Planet of the Apes. http://www.outsidethebeltway.com/dirty_apes/

Kaya, N., & Erkip, F. (1999). Invasion of Personal Space Under the Condition of Short-Term Crowding: A Case Study on an Automated Teller Machine. *Journal of Environmental Psychology*, 19, 183–189.

Kedmey, D. (2015, Feb. 9). Virtually Real: Microsoft Joins the Crowd Betting on 3-D Headsets. *Time Magazine*. p. 12.

Kedmey, D. (2015, May 25). Caught by the Cloud: Will Better Software Help Police Put Criminals behind Bars? *Time Magazine*, p. 18.

Keefe, D. F. (2011). From Gesture to Form: The Evolution of Expressive Freehand Spatial Interfaces. *Leonardo, 44*(5), 460–461.

Kellogg, C. (2014, June 3). Amazon and Hachette: The Dispute in 13 Easy Steps. *Los Angeles Times*. http://www.latimes.com/books/jacketcopy/la-et-jc-amazon-and-hachette-explained-20140602-story.html#page=1

Knowles, E. (1972). Boundaries around Social Space: Dyadic Responses to an Invader. *Environment and Behavior, 4*(4), 437–445.

Knowles, E. S., Kreuser, B., Hass, S., Hyde, M., & Schuchart, G. E. (1976). Group Size and the Extension of Social Space Boundaries. *Journal of Personality and Social Psychology, 33*(5), 647–654.

Koskela, H. (2003). 'Cam Era'—the Contemporary Urban Panopticon. *Surveillance and Society, 1*(3), 292–313.

Kramer, A. D. I., Guillory, J. E., & Hancock, J. T. (2014). Experimental Evidence of Massive-Scale Emotional Contagion through Social Networks. *Proceedings of the National Academy of Science*.

Kratochwill, L. (2015, March 16). SXSW 2015: How Extreme Bionics Will Rid the World of Disability—Robotic Limbs Will Become an Advantage, Not an Impediment. *Popular Science*. http://www.popsci.com/sxsw-2015-how-extreme-bionics-will-help-rid-world-disability

Kress, G., & van Leeuwen, T. (2001). *Multimodal Discourse: The Modes and Media of Contemporary Communication*. London: Hodder-Arnold.

Lane, N., Pengyu, L., Zhou, L., & Zhao, F. (2014). Connecting Personal-Scale Sensing and Networked Community Behavior to Infer Human Activities. *UbiComp '14 Adjunct*, 595–606.

Laney, D. (2001, Feb. 6). 3D Data Management: Controlling Data Volume, Velocity, and Variety. META Group Application Delivery Strategies. http://blogs.gartner.com/doug-laney/files/2012/01/ad949-3D-Data-Management-Controlling-Data-Volume-Velocity-and-Variety.pdf

Lefebvre, H. (2003). *The Urban Revolution* (R. Bononno, Trans.). Minneapolis, MN: University of Minnesota Press. (originally published in French in 1970)

Lindstrom, M. (2011, Nov. 7). Zones of Seduction. How Supermarkets Turn Shoppers into Hoarders. *Time Magazine*, p. 58.

Lin, M. (2000). *Boundaries*. New York: Simon & Schuster.

Liu, Y., Goncalves, J., Ferreira, D., Hosio, S., & Kostakos, V. (2014). Identity Crisis of Ubicomp? Mapping 15 Years of the Field's Development and Paradigm Change. *Ubicomp '14 Adjunct*, 75–86.

Lv, M., Chen, L., Chen, G., & Zhang, D. (2014). Detecting Traffic Congestions Using Cell Phone Accelerometers. *UbiComp '14 Adjunct*, 107–110.

Madden, M., & Rainie, L. (2015, May 20). Americans' Attitudes about Privacy, Security and Surveillance. *Pew Research Center Reports*. http://pewinternet.org/2015/05/20/americans-attitudes-about-privacy-security-and-surveillance/

Marquardt, N. & Greenberg, S. (2012, June–Aug). Informing the Design of Proxemic Interactions. *Pervasive Computing*, 14–23.

Mars, R. (2012, March 9). Queue Theory and Design. 99% *invisible*, Ep. 49. http://99percent-invisible.org/transcript-49-queue-theory-and-design/

Mars, R. (2014, Feb. 25). U.T.B.A.P.H. 99% *Invisible*, 103. http://99percentinvisible.org/episode/u-t-b-a-p-h/

Mars, R. (2014, March 11). One Man Is an Island. 99% *Invisible (Episode 105)*. http://99per-centinvisible.org/episode/one-man-is-an-island/

Marwick, A. (2012). The Public Domain: Social Surveillance in Everyday Life. *Surveillance & Society*, 9(4), 378–393.

Marwick, A. E., & boyd, d. (2010). I Tweet Honestly, I Tweet Passionately: Twitter Users, Context Collapse, and the Imagined Audience. *New Media & Society*, 13(1), 114–133.

Matic, A., Osmani, V., & Mayora-Ibarra, O. (2012). Analysis of Social Interactions through Mobile Phones. *Mobile Network Applications*, 17, 808–819.

Mattern, S. (2014, April). Interfacing Urban Intelligence. *Places Journal*. https://placesjournal.org/article/interfacing-urban-intelligence/

Matsuda, Y., & Arai, I. (2014). A Safety Assessment System for Sidewalks at Night Utilizing Smartphones' Light Sensors. *UbiComp '14 Adjunct*, 115–118.

McArthur, J. A. (2009). Digital Subculture: A Geek Meaning of Style. *Journal of Communication Inquiry*, 33(1), 58–70.

McArthur, J. A., & Bostedo-Conway, K. (2012). Exploring the Relationship between Student-Instructor Interaction on Twitter and Student Perceptions of Teacher Behaviors. *International Journal of Teaching and Learning in Higher Education*, 23(3), 286–292.

McAvinney, P. (2014). TEDxGreenville talk. http://tedxgreenville.com/portfolio/2014-paul-mcavinney

McCarthy, E. (2011, July 7). Andy Serkis & the Evolution of Performance Capture Tech. *Popular Mechanics*. Retrieved from http://www.popularmechanics.com/technology/digital/visual-effects/andy-serkis-and-the-evolution-of-performance-capture-tech

McLuhan, M. (1962). *The Gutenberg Galaxy: The Making of Typographic Man*. University of Toronto Press.

McLuhan, M. (1964). *Understanding Media: The Extensions of Man*. Chicago: McGraw-Hill.

McLuhan, M. (1967). *The Medium Is the MASSAGE*. New York: Random House–Bantam Books

Meyrowitz, J. (1985). *No Sense of Place: The Impact of Electronic Media on Social Behavior*. New York: Oxford University Press.

Mitchell, W. J. T. (2012). Image, Space, Revolution: The Arts of Occupation. *Critical Inquiry*, 39(1), 8–32.

Monahan, T. (2015). The Right to Hide? Anti-surveillance Camouflage and the Aestheticization of Resistance. *Communication and Critical/Cultural Studies*, 12(2), 159–178.

Montello, D. R., & Sas, C. (2006). Human Factors of Wayfinding in Navigation. In *International Encyclopedia of Ergonomics and Human Factors*. Boca Raton, FL: CRC Press/Taylor & Francis, Ltd., pp. 2003–2008.

Newitz, A. (2015, April 1). The Guerrilla Movement to Deploy Sensors All Over Your City. *GrayArea.org (Gray Area Foundation for the Arts)*. http://grayarea.org/press/the-guerrilla-movement-to-deploy-sensors-all-over-your-city/

Norman, D.A. (2004). *Emotional Design: Why We Love (or Hate) Everyday Things*. New York: Basic Books.

Oldenburg, R. (1999). *The Great Good Place: Cafes, Coffee Shops, Bookstores, Bars, Hair Salons, and Other Hangouts at the Heart of a Community*. New York, NY: Marlowe & Company.

Osmond, H. (1957). Function as the Basis of Psychiatric Ward Design. *Psychiatric Services* 8(4), 23–29.

Oswalt, P. (2014, Sept. 8–15). Why I Quit Twitter—and Will Again. *Time Magazine*. p. 28.

Pace, C. R., & Stern, G. G. (1958). An Approach to the Measurement of Psychological Characteristics of College Environments. *Journal of Educational Psychology, 49*, 269–277.

Page, L. (2015, Aug. 10). Google Announces Plan for New Operating Structure. https://investor.google.com/releases/2015/0810.html

Pariser, E. (2011). *The Filter Bubble: What the Internet Is Hiding from You*. New York: The Penguin Press.

Park, K., Shin, H., & Cha, H. (2013). Smartphone-based pedestrian tracking in indoor corridor environments. *Perspectives on Ubiquitous Computing, 17*, 359–370.

Patterson, M. L., Lammers, V. M., & Tubbs, M. E. (2014). Busy Signal: Effects of Mobile Device Usage on Pedestrian Encounters. *Journal of Nonverbal Behavior, 38*(3), 313–324.

Perera, C., Liu, H., & Jayawardena, S. (2015). The Emerging Internet of Things Marketplace from an Industrial Perspective: A Survey. *Emerging Topics in Computing, IEEE Transactions*. Available in pre-print: http://ieeexplore.ieee.org/xpl/articleDetails.jsp?reload=true&arnumber=7004800

Poeschel, S., & Doring, N. (2010). Nonverbal Cues in Mobile Phone Text Messages: The Effects of Chronemics and Proxemics. In R. Ling & S. Campbell (Eds.), *The Reconstruction of Space and Time: Mobile Communication Practices*. New Brunswick, NJ: Transaction Publishers.

Poldma, T. (2010). Transforming Interior Spaces: Enriching Subjective Experiences through Design Research. *Journal of Research Practice, 6*(2), 1–12.

Poldma, T., & Wesolkowska, M. (2005). Globalisation and Changing Conceptions of Time-Space: A Paradigm Shift for Interior Design? *In Form, 5*, 54–61.

Prensky, M. (2001). Digital Natives, Digital Immigrants. *On the Horizon, 9*(5), 1–6.

Preston, P. (2005). Proxemics in Clinical and Administrative Settings. *Journal of Healthcare Management, 50*(3), 151–154.

Rainie, L., & Madden, M. (2015, March 16). Americans' Privacy Strategies Post-Snowden. *Pew Research Center Reports*. http://www.pewinternet.org/2015/03/16/americans-privacy-strategies-post-snowden/

Rashid, Z., Pous, R., Melia-Segui, J., & Morenza-Cinos, M. (2014). Mobile Augmented Reality for Browsing Physical Spaces. *Ubicomp '14 Adjunct*, 155–158.

Rashid, Z., Pous, R., Melia-Segui, J., & Peig, E. (2014). Cricking: Browsing Physical Space with Smart Glass. *Ubicomp '14 Adjunct*, 151–154.

Rautaray, S. S., & Agrawal, A. (2015). Vision Based Hand Gesture Recognition for Human Computer Interaction: A Survey. *Artificial Intelligence Review, 43*, 1–54.

Remland, M. S., Jones, T. S., & Brinkman, H. (1995). Interpersonal Distance, Body Orientation, and Touch: Effects of Culture, Gender, and Age. *The Journal of Social Psychology, 135*(3), 281–297.

Rheingold, H. (2002). *Smart Mobs: The Next Social Revolution*. Cambridge, MA: Basic Books.

Rheingold, H. (2010). Attention, and Other 21st-Century Social Media Literacies. *EDUCAUSE Review, 45*(5), 14–24.

Riley v. California, 573 U. S. ____ (2014).

Ring, B. (2015, Aug. 13). Zero Latency: The VR Revolution Begins in Melbourne, Australia. *CNET*. http://www.cnet.com/news/zero-latency-vr-entertainment-revolution-begins-melbourne-australia/

Rochman, B. (2010, May 3). Take Two Texts and Call me in the Morning: How a New Phone-based Initiative Seeks to Improve Prenatal Care. *Time*, p. 54.

Roschke, K. (2014). Passionate Affinity Spaces on Reddit.com: Learning Practices in the Gluten-Free Subreddit. *Journal of Digital and Media Literacy, 2*(1). http://www.jodml.org/2014/06/17/passionate-affinity-spaces-on-reddit-com-learning-practices-in-the-gluten-free-subreddit/

Rodriguez Garzon, S., & Deva, B. (2014). Geofencing 2.0: Taking Location Based Notifications to the Next Level. *Ubicomp '14 Adjunct*, 921–932.

Sarasohn-Kahn, J. (2008). *The Wisdom of Patients: Healthcare Meets Online Social Media*. Oakland: California HealthCare Foundation.

Self, W. (2015, January 30). Does Technology Make People Touch Each Other Less? *A Point of View*. BBC radio broadcast. http://www.bbc.com/news/magazine-31026410

Setti, F., Russell, C., Bassetti, C., & Cristani, M. (2015). F-Formation Detection: Individuating Free-Standing Conversational Groups in Images. *PLoS ONE 10*(5), 1–26.

Shedroff, N. (2001). *Experience Design 1*. Indianapolis, IN: New Riders Publishing.

Singh, S. (1999). *The Code Book: The Science of Secrecy from Ancient Egypt to Quantum Cryptography*. New York: Anchor Books.

Smith, R. (2013). Mediated Competition, Comparison, and Connections: How Mobile Interactive Fitness Technologies Alter the Cycling Experience. Paper presented at the National Communication Association Annual Conference, Washington, DC.

Sommer, R. (1969) *Personal Space: The Behavioral Basis of Design*. Englewood Cliffs, NJ: Prentice-Hall, Inc.

Spohrer, J. (1999). Information in Places. *IBM Systems Journal, 38*(4), 602–628.

Stafford-Frasier, Q. (1995). The Trojan Room Coffee Pot: A (Non-Technical) Biography. http://www.cl.cam.ac.uk/coffee/qsf.html

Stafford-Frasier, Q. (2001). When Convenience Was the Mother of Invention. *Communications of the Association for Computing Machinery, 44* (7), 25.

Stein, J. (2012, April 23). Don't Speak, Memory. If the Information Age Doesn't Shut Up Soon, All Our Best Stories Will Be Ruined. *Time Magazine*. p. 62.

Stein, J. (2015, Aug. 17). Inside the Box. *Time Magazine*. pp. 40–49.

Stenger, C. (2014). The Great Smartphone Debate, Simply Stated Blog, *Real Simple*. http://simplystated.realsimple.com/2014/07/09/the-great-smartphone-debate/

Steuer, J. (1992). Defining Virtual Reality: Dimensions Determining Telepresence. *Journal of Communication*, 4, 73–93.

Stodghill, R. (2015). *Where Everybody Looks Like Me: At the Crossroads of America's Black Colleges and Culture*. New York: Harper Collins.

Stole, I. L. (2014). Persistent Pursuit of Personal Information: A Historical Perspective on Digital Advertising Strategies. *Critical Studies in Media Communication*, 31(2), 129–133.

Straker, L. M., O'Sullivan, P. B., Smith, A., & Perry, M. (2007). Computer Use and Habitual Spinal Posture in Australian Adolescents. *Public Health Reports*, 111(5), 634–643.

Strange, C. C., & Banning, J. H. (2001). *Educating by Design: Creating Campus Learning Environments That Work*. San Francisco: Jossey-Bass.

Sukale, R., Voida, S., & Koval, O. (2014). The Proxemic Web: Designing for Proxemic Interactions with Responsive Web Design. *Ubicomp' 14 Adjunct*, 171–174.

Troester, M. (2012). Big Data Meets Big Data Analytics [White Paper]. Cary, NC: SAS Institute, Inc. http://www.sas.com/content/dam/SAS/en_us/doc/whitepaper1/big-data-meets-big-data-analytics-105777.pdf

Turbow, J. (2012, August 3). Bigger, Faster, Stronger. Will Bionic Limbs Put the Olympics to Shame? *Wired*. http://www.wired.com/2012/08/next-gen-prosthetics-and-sports/

Turkle, S. (2011). *Alone Together: Why We Expect More from Technology and Less from Each Other*. New York: Basic Books.

United States Holocaust Memorial Museum. (2006). Unites States Holocaust Memorial Museum. http://www.ushmm.org

Vermesan, O., & Friess, P. (Eds.). (2014). *Internet of Things—From Research and Innovation to Market Deployment*. Aalborg, DK: River Publishers.

Wæraas, A. (2009). On Weber. In Ø. Ihlen, B. van Ruler, & M. Fredriksson (Eds.), *Public Relations and Social Theory : Key Figures and Concepts* (pp. 301–322). New York: Routledge.

Walsh, B. (2014, Nov. 24). Data Mine: The Next Revolution in Personal Health May Be the Little Step-Tracking Band on Your Wrist. *Time*. pp. 34–38.

Walsh, W. B. (1973). *Theories of Person-Environment Interaction: Implications for the College Student*. Iowa City, IA: American College Testing Program.

Walther, J. B. (1992). Interpersonal Effects in Computer-Mediated Interaction: A Relational Perspective. *Communication Research*, 19, 52–90.

Walther, J. B. (1996). Computer-Mediated Communication: Impersonal, Interpersonal, and Hyperpersonal Interaction. *Communication Research*, 23, 3–43.

Walther, J. (2006). Nonverbal Dynamics in Computer Mediated Communication, or :(and the Net :('s with You, :) and You :) Alone. In V. Manusov & M. Patterson (Eds.), *The Sage Handbook of Nonverbal Communication* (pp. 461–480). Thousand Oaks, CA: Sage.

Watson, M. D., & Atuick, E. A. (2015). Cell Phones and Alienation among Bulsa of Ghana's Upper East Region: "The Call Calls You Way." *African Studies Review*, 58(1), 113–132.

Wayne, T. (2015, Feb. 20). Shutterbug Parents and Overexposed Lives. *The New York Times*. http://www.nytimes.com/2015/02/22/style/shutterbug-parents-and-overexposed-lives.html

Weisgerber, C., & Butler, S. (2011). Social Media as a Professional Development Tool: Using Blogs, Microblogs and Social Bookmarks to Create Personal Learning Networks. In C. Wankel (Ed.), *Teaching Arts & Science with Social Media*. Bingley, U.K.: Emerald.

White, A. F., & McArthur, J. A. (forthcoming). Twitter Chats as Third Places: Conceptualizing a Digital Gathering Site. *Social Media + Society*.

Wilken, R., & Goggin, G. (2012). *Mobile Technology and Place*. New York: Routledge.

Wilken, R., & Goggin, G. (2014). *Locative Media*. New York: Routledge.

Williams, D. (2010). The Mapping Principle, and a Research Framework for Virtual Worlds. *Communication Theory, 20*, 451–470.

Williams, S. (2007). User Experience Design for Technical Communication: Expanding Our Notions of Quality Information Design. *Proceedings of the Annual Meeting of the IEEE Professional Communication Society*.

Young, J. G., Trudeau, M. B., Odell, D., Marinelli, K., & Dennerlein, J. T. (2013). Wrist and Shoulder Posture and Muscle Activity during Touch-Screen Tablet Use: Effects of Usage Configuration, Tablet Type, and Interacting Hand. *Work, 45*(1), 59–71.

Yuqing, R., Harper, F., Drenner, S., Terveen, L., Kiesler, S., Riedl, J., & Kraut, R. E. (2012). Building Member Attachment in Online Communities: Applying Theories of Group Identity and Interpersonal Bonds. *MIS Quarterly, 36*(3), 841–864.

Zickuhr, K. (2013, Sept. 12). Location-Based Services. Pew Research Center. http://www.pew-internet.org/2013/09/12/location-based-services/

ABOUT THE AUTHOR

John A. McArthur is an associate professor in the James L. Knight School of Communication at Queens University of Charlotte. He currently serves as director of graduate programs and was instrumental in developing the $5.75 million grant awarded to the school by the John S. & James L. Knight Foundation in 2010. He previously served as director of undergraduate programs in the school and has worked with faculty university-wide as the director of online faculty services, a faculty development role promoting successful teaching and learning in online and hybrid environments.

Dr. McArthur received his Ph.D. in rhetorics, communication, and information design from Clemson University. His research and teaching focuses on proxemics (the use of space) and information design with a particular interest in the ways that digital technology influences our interactions with spaces. His studies have applied information design theories to examine both built and digital spaces—urban settings, chat rooms, memorials, public art, classroom designs, and campus architecture. Some of his research also focuses on the relationships among spaces of learning, technology, and successful instructional practice. His research can be found in journals ranging from the *Journal of*

Communication Inquiry, Communication Teacher, and the *International Journal of Teaching and Learning in Higher Education,* to the *American Clearinghouse of Educational Facilities Journal, Journal of Library Innovation,* and the *Journal of Learning Spaces.*

INDEX

General Editor: *Steve Jones*

Digital Formations is the best source for critical, well-written books about digital technologies and modern life. Books in the series break new ground by emphasizing multiple methodological and theoretical approaches to deeply probe the formation and reformation of lived experience as it is refracted through digital interaction. Each volume in **Digital Formations** pushes forward our understanding of the intersections, and corresponding implications, between digital technologies and everyday life. The series examines broad issues in realms such as digital culture, electronic commerce, law, politics and governance, gender, the Internet, race, art, health and medicine, and education. The series emphasizes critical studies in the context of emergent and existing digital technologies.

Other recent titles include:

Felicia Wu Song
  *Virtual Communities: Bowling Alone, Online
  Together*

Edited by Sharon Kleinman
  *The Culture of Efficiency: Technology in
  Everyday Life*

Edward Lee Lamoureux, Steven L. Baron, &
  Claire Stewart
  *Intellectual Property Law and Interactive
  Media: Free for a Fee*

Edited by Adrienne Russell & Nabil Echchaibi
  *International Blogging: Identity, Politics and
  Networked Publics*

Edited by Don Heider
  Living Virtually: Researching New Worlds

Edited by Judith Burnett, Peter Senker
  & Kathy Walker
  *The Myths of Technology:
  Innovation and Inequality*

Edited by Knut Lundby
  *Digital Storytelling, Mediatized
  Stories: Self-representations in New
  Media*

Theresa M. Senft
  *Camgirls: Celebrity and Community
  in the Age of Social Networks*

Edited by Chris Paterson & David
  Domingo
  *Making Online News: The
  Ethnography of New Media
  Production*

To order other books in this series please contact our Customer Service Department:
(800) 770-LANG (within the US)
(212) 647-7706 (outside the US)
(212) 647-7707 FAX

To find out more about the series or browse a full list of titles, please visit our website:
WWW.PETERLANG.COM